Contents

Quarter-scale Fokker Triplane by Ian Whiting displays two modern trends – scale and large size.

Introducing Radio Control MODEL AIRCRAFT

Bill Burkinshaw

ARGUS BOOKS

Argus Books
Argus House
Boundary Way
Hemel Hempstead
Hertfordshire HP2 7ST

First published in 1983
Reprinted 1984, 1986, 1987, 1988, 1989
© Bill Burkinshaw

ISBN 0 85242 801 4

Set by SIOS Ltd. 111-115 Salisbury Road,
London NW6 6RJ
Printed and bound by LR Printing Services Ltd, Edward Way,
Burgess Hill, West Sussex, RH15 9UA, England

CHAPTER ONE

Introduction

Modern radio control (R/C) can provide aircraft enthusiasts with pleasure and satisfaction at many levels. For those who enjoy building intricate scale models detailed to the last screw and rivet, it is the means to achieve the final step to realism in miniature, from taxi out to the runway, through take-off, intricate manoeuvres, precise aerobatics, and back to earth for a smooth landing. Others view the creative building and finishing process as a necessary evil, which interrupts the progress between purchase and flying site, gaining their pleasure from flying as a sport, seeing little connection between building the model and flying it, in much the same way as, say, a footballer might view his football.

There are, of course, all the shades in between these two extremes of enthusiasm, all of them enjoying one of the pleasures that up-to-date electronic technology can offer. It should be pointed out immediately that modern R/C is for everyone, no need to be an electronic wizard to use it, remember we are in the age of plug-in miracles, just add the batteries, switch on and go! No need even to add the batteries to many up-to-the-minute R/C systems, for with rechargeable battery packs and mains electricity-operated chargers, the order is plug in, charge up, switch on and go. As with the family car, there are R/C systems to match every would-be enthusiast's budget, and almost month by month the effects of technological development reduce prices as the increasing market for R/C equipment allows manufacturers to take advantage of economies of scale.

Typical club flying site scene with modellers grouped around the flying patch. Several models airborne with pilots gazing intently skywards as another prepares to take off.

Maybe the thought of an internal combustion engine puts you off, memories of an awkward recalcitrant beast of an engine from years ago? Firstly it is fair to say that the modern model aircraft IC power plant starts more readily, runs more smoothly, more quietly and produces more power than its counterpart of 20 years ago. Glowplug ignition engines are almost universal, as are electric starters. Still not convinced? Perhaps silent flight, or almost silent flight in the case of electric power, might be the answer. Gliders of various types, suited for several different types of operation, are very popular, probably accounting for one third of R/C aircraft interest. R/C gliders are flown from flat-field sites, where modellers seek to prolong flights by using thermal upcurrents of air, or from windward slopes where the models ride the wind sweeping up the slope faces. This latter activity demands that the modeller is exposed to some of the most attractive scenery in the country and develops a strong interest in hiking, as the models are certainly not launched from the foot of those windward slopes!

Soarers

Flat-field gliders, or thermal soarers, can be launched by a variety of methods (described in detail in Chapter 8). All methods are based on the kiting principle; the model is towed up on a long line to an altitude of up to 150 metres, released, and then hopefully flown by its pilot into a rising current of warm air known as a thermal. A flight duration which would be as short as 3-4 minutes in still air can be boosted to 15-20 minutes or even longer – flights of an hour are not uncommon for a glider riding thermal currents.

For the slope-soaring enthusiast the length of flight is dictated by the direction and strength of the wind. While the wind flows on to the slope face and is strong enough, and not too strong, the model will continue to fly. At the time of writing, the world record for slope-soaring flight is in excess of 32 hours.

Powered models

For those whose interests are devoted to the noisier side of life, and there is undoubted fascination in the operation of small high-performance IC engines, there is a wide range of R/C activity for them to choose from. Pure sport flying, simply taking off, flying round and landing, is sufficient achievement for some. Many enjoy the challenge of aerobatics: loops, rolls, spins and inverted flying and the many combinations of these basic aerobatic manoeuvres tax the concentration and reflexes, as does the racing of up to four high-speed model racers round a triangular course, or pylon racing, as it is known. Scale models appeal to most R/C modellers, though not all of them have the patience and long-term interest in model construction to build such models, however.

It would not be right to leave any outline of powered model types without mention of helicopters and autogyros. They are not the same, it should be realised, for while both are supported by rotating wings (the rotors), in the case of the helicopter the rotors are driven round by an engine whilst the rotation of the autogyro's rotor is caused by the machine's forward motion through the air. An

Most trainer models are of the high wing layout. This large Japanese kit model is powered by a 4 stroke engine.

autogyro cannot normally take-off vertically, although some examples with power take-off from the engine to spin up the rotor very nearly achieve vertical lift-off.

A true helicopter can lift off the ground vertically and hover over a fixed point, move sideways, backwards, forwards – in fact in any direction in three dimensions. The best of model helicopters can duplicate fully the performance of the full-size, and even better, being fully aerobatic, able to loop-the-loop, roll and even fly upside down. The latter manoeuvre is really only a party trick, not a very useful accomplishment! Radio control helicopters are by far the most challenging type of model aircraft to control, since hovering flight requires simultaneous operation of four controls, each one vital to the stability of the model. There is one school of thought that argues that for a novice R/C pilot, a helicopter is no more difficult to learn to control than a conventional fixed-wing aircraft, reasoning that it is the experienced pilots' very familiarity with conventional model aircraft control that makes adaptation to helicopter flying difficult. Without direct experience, it would be unwise of me to encourage others to adopt this course, suffice it to say that helicopters are difficult to learn to fly.

Controlling the aircraft

So far no mention has been made of what the radio is actually used to control. I have mentioned take-offs, aerobatics, racing, landings etc. but without any explanation of the mechanics of the operation. Model aircraft function in exactly the same medium as real aeroplanes, and they are subject to the same physical laws such as gravity, centrifugal force, deceleration etc. It follows that control of the aircraft will be effected in a like manner.

Aerodynamics, the study of objects in moving currents of air (aeroplanes) is a complex subject beyond the scope of this book, but some basic knowledge of the subject certainly helps in understanding the effects of such things as balance,

9

behaviour in turns, and effects of control surfaces. Lack of this knowledge need not hinder the would-be enthusiast, but for many, increases interest and pleasure.

A very brief outline of the forces affecting an aircraft in flight is essential for the purposes of explanation of control surface action. An aircraft in stable, level flight at a constant speed is subject to a combination of balanced forces. Thrust or power moves the aircraft forward, opposed by drag, the air resistance. Power must exceed drag for the aircraft to increase speed. Lift, generated by the wings, must

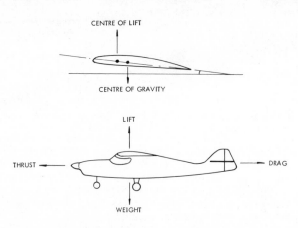

Fig. 1

equal the mass of the aircraft for flight, and exceed it for the aircraft to gain height. The basic unadorned wing of an aircraft is unstable and will tend to rotate if it is not stabilised in some way; in conventional aircraft a tailplane (or in American terminology a stabiliser) prevents this from happening (see fig 2). Stability in a fore and aft direction is determined by the action of the tailplane, but lateral

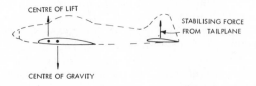

Fig. 2

stability, necessary to prevent the aircraft flipping over on to its back, is determined by angling the wings to one another, dihedral angle (see fig 3), pendulum stability, wing sweepback, or a combination of all three.

Controlling the aircraft basically requires that this stability is destroyed in a controlled manner. This is achieved by increasing or decreasing engine power, creating a lift differential from one wing to the other by movable control surfaces on the wings called ailerons, upsetting the stabilising effect of the tailplane by movable surfaces called elevators, and finally influencing the directional stability by use of a rudder.

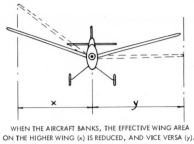

WHEN THE AIRCRAFT BANKS, THE EFFECTIVE WING AREA
ON THE HIGHER WING (x) IS REDUCED, AND VICE VERSA (y).
THE 'LARGER' WING (y) LIFTS MORE AND RIGHTS THE AIRCRAFT

Fig. 3

All of these controls act by changing the aerodynamic lifting characteristics of the control surface in question. Consider the action of the tailplane first of all. Normally the tailplane generates sufficient lifting force to balance exactly the rotating tendency of the wing. If the lifting force of the tailplane is increased by depressing the elevator, the tailplane will rise in relation to the wing, depressing the nose of the aircraft, which dives. Raise the elevator and, providing there is sufficient power available, the aircraft climbs. If there is enough excess power available, the climb becomes a loop, one of the basic aerobatic manoeuvres. Rolling or rotation about the the longitudinal axis is effected in a like manner, except that two control surfaces, ailerons, one on each wing, are involved. When one aileron is depressed, increasing the lift of the wing, the opposing aileron is raised, decreasing the lift on that wing. To be strictly correct, it increases the lift in a downward direction! The result is a roll towards the raised aileron which if continued results in the aircraft completely rotating about its axis. Although the fin and rudder are vertical, in aerodynamic terms the rudder moving in relation to the fin increases lift either to one side or the other, causing the aircraft to yaw. The effects of these control surface movements are shown in fig 5.

All of these control surfaces can be moved by radio control, and the modeller can therefore have complete control over the behaviour of the model in the air. Actual movement of the control surfaces is done by miniature motorised actuators known as servos. Electric motors provide the power, coupled via reduction gearboxes to the surfaces. Servos are put into operation by command signals received by the model R/C receiver from the model flier's transmitter. As well as the primary aerodynamic controls, secondary controls such as retracting wheels, airbrakes, bomb dropping, camera operation and others are all possible.

For control of gliders, the basic minimum number of controls would be two, rudder and elevators. These two controls are quite sufficient for full control over the model, providing the facility for quite a range of aerobatic manoeuvres, providing that the design of the model allows. More advanced gliders usually include aileron control, sometimes as an alternative to rudder, but for full control, particularly for aerobatics, the full complement of rudder, elevator and ailerons is necessary. Many high-performance gliders feature spoilers operated by the R/C system. These do just as their name implies, spoil the efficiency of the wings, enabling the pilot to control the rate of descent of the model more precisely when landing or descending from heights.

Power-flying enthusiasts almost certainly need one additional control function

11

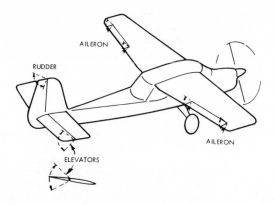

AILERON

RUDDER

AILERON

ELEVATORS

Fig. 4

right from the outset of their activity, the ability to control engine speed. It is far more desirable to have the availability of power in excess in the model, for with the R/C system, the pilot is able to select just the right amount of power for any situation, full power for take-off, cruise power and power off for landing. It is certainly not a good idea to fit an engine with too little, or marginal power output, for if the engine for any reason delivers less than its best, the resulting situation can be disastrous, particularly for the novice. Rudder, elevator, motor ("REM") is an almost universal concept for the novice R/C pilot, who can then progress to a model including aileron control. It might be asked why aileron control is not normally included as a control function for R/C novices' models'. In broad terms the requirements for a first R/C model aircraft, be it glider or power, include a large degree of inherent stability. This generally implies that the previously mentioned dihedral angle should be sufficient to provide a strong self-righting force to the model, so that if the novice does get into difficulties while flying, he can let the model sort itself out without interference. Unhappily, although ailerons provide a more positive control action in starting turns, and correcting unwanted turns, they do not operate very efficiently when the wings are set at a generous dihedral angle as dictated by the demand for stability. Taken to either extreme, no dihedral, or a great deal, mean that in the former case the model will have virtually no self-righting ability, not the ideal for a learner, or in the latter so much stability as to render ailerons virtually ineffective.

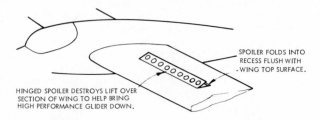

SPOILER FOLDS INTO
RECESS FLUSH WITH
.WING TOP SURFACE.

HINGED SPOILER DESTROYS LIFT OVER
SECTION OF WING TO HELP BRING
HIGH PERFORMANCE GLIDER DOWN.

Fig. 5

12

ANGLE OF ATTACK OF WINGS INCREASED,
LIFT INCREASES, AIRCRAFT CLIMBS

ELEVATOR UP CAUSES TAIL
TO BE DEPRESSED

ANGLE OF ATTACK REDUCED, WINGS
LIFT LESS, AIRCRAFT DIVES

ELEVATOR DOWN CAUSES
TAIL TO RISE

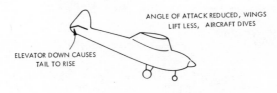

Fig. 5a

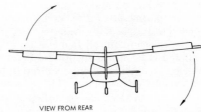

VIEW FROM REAR

RIGHT AILERON UP, LEFT DOWN AND
AIRCRAFT BANKS TO THE RIGHT

Fig. 5b

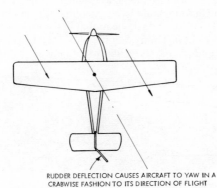

RUDDER DEFLECTION CAUSES AIRCRAFT TO YAW IN A
CRABWISE FASHION TO ITS DIRECTION OF FLIGHT

Fig. 5c

13

Former World Scale Champion Mick Reeves wheels out his Spitfire.

CHAPTER TWO

Radio Control Equipment

So much for the basics of R/C aircraft modelling. Now for more detail. It may be stating the obvious but, nonetheless, for the record, any Radio or RF (Radio Frequency) link requires the presence of a Transmitter and Receiver. In addition to these two items, radio control requires some means of translating commands transmitted and received into control surface movement in the aircraft model to be controlled. This final link in the R/C chain is called a Servo.

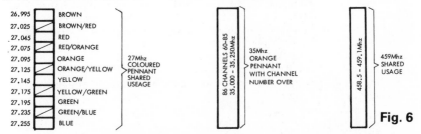

Fig. 6

The performance of the equipment almost always gives rise to queries from the novice, when it is first encountered. Simple facts such as "How far away does it work?" "How can more than one model be operated at a time?" are the obvious puzzlers. Rest assured that all modern R/C equipment has adequate range. Quoting range in terms of metres or kilometres can be misleading – a simpler description such as 'Out of Sight Range' is more meaningful, for it is obvious that unless the model can be adequately seen with the naked eye, it will not be possible to control it properly. This definition of range covers the situation well, for many variables govern the finite range of the system. In simple terms range is a function of the output power of the transmitter, the sensitivity of the receiver and local conditions prevailing.

As with any other radio frequency transmitting device, model R/C equipment is subject to various regulations governing its use. Most national governments have strict laws on this subject, and in the UK the Home Office administers RF transmission under the Wireless Telegraphy Acts, which lay down specific operating frequencies for model control transmitters. In the UK the frequencies available for model aircraft use are 26.960-27.280MHz; 35.000-35.250MHz and 458.5-458.8MHz. Within these broad bands of frequency allocation there are numerous individual channels or spot frequencies available. Because many modellers may wish to operate simultaneously, the transmitters and receivers used are all "crystal controlled". The system of crystal control ensures that each transmit-

15

ter only operates on the exact frequency required, and does not interfere with other users. A system of coloured and numbered pennants is used to enable operators to make a quick visual check on which "spots" are in use before switching on their own equipment. Simultaneous use of a single spot frequency by more than one operator is not practical. If two transmitters are switched on on the same channel, one or both models being flown will crash. It is usually possible for the user to change from one spot to another by simply unplugging the crystals from the transmitter and receiver, and substituting an alternative matched pair of crystals for an unused channel. Most modellers have several pairs of crystals available and use whichever frequency is available to them.

Changing from one of the frequency *bands* to another is not quite so simple. The widely differing designs needed for transmitting at the different frequency means that much more of an alteration to the equipment is necessary and just changing crystals is not enough. Many of the "top-of-the-line" models of current R/C equipment manufacturers do feature the possibility of changing from one frequency to another, achieved by changing a complete section of the transmitter and receiver, usually known as a module, allowing the operator to change at will from 27MHz to 35MHz operation. Changing to 459MHz is more complicated, but some modular equipment is available that can be used with 459MHz modules. It should be pointed out that modules need to be changed on both transmitter and receiver. Once the modules for any particular band have been fitted, it is also possible to change crystals so that any spot frequency within the band can be used.

Having put the basic question to one side, let us look at the various elements of the system in more detail.

Transmitter

For the purposes of description, a four-function system suitable for full control of a powered model aircraft will be described. The most obvious features of the transmitter are usually the two control systems (joysticks) placed on either side of the case, conveniently placed for operation, one by each hand. These sticks are dual axis and can be moved in both axes simultaneously. Two control configurations are used as follows: Mode 1: Right-hand stick controls primary steering

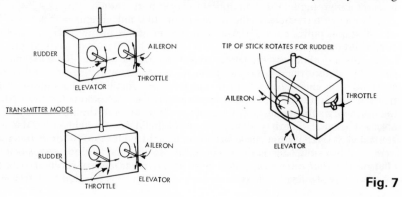

Fig. 7

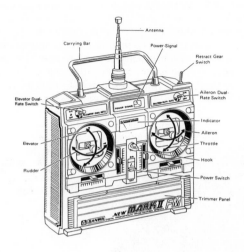

Labels on figure:
- Antenna
- Carrying Bar
- Power-Signal
- Retract Gear Switch
- Elevator Dual-Rate Switch
- Aileron Dual-Rate Switch
- Indicator
- Aileron
- Elevator
- Throttle
- Hook
- Rudder
- Power Switch
- Trimmer Panel

SANWA *NEW* MARK II FM

Fig. 8

function/roll control (ailerons in a four-function model, rudder if only three servos are used) and throttle (capable of being placed in any position by either a ratchet or friction system). Left-hand controls elevator and rudder (secondary steering or yaw control in a four-function model), both self-centring. Mode 2: Right-hand stick controls primary steering and elevator, both self-centring. Left-hand stick controls throttle (ratchet or friction) and rudder self-centring. It is, of course, possible to use any number of other arrangements, but beware, failure to adhere to one of the accepted modes may make it impossible for you to avail yourself of expert advice and help, as few even experienced R/C pilots will attempt to fly a model using an unfamiliar transmitter mode.

Convention also dictates the direction of stick movements as follows – Up elevator, stick pulled back or towards the bottom of the case, and vice versa. Right aileron, stick moved to the right (N.B. for 'Right' aileron movement the RH aileron should go up, and the left down, when viewed from the rear of the model). Throttle, full throttle with the stick pushed forward or towards the top of the case.

Allied with each of the control stick axes are small subsidiary levers often marked with graduations. These levers are known as "trims" and their function is to adjust the neutral position of the function. Why this is desirable is quite simple. Should the model have a tendency to turn continually to the right, either the main control stick will have to be held with a permanent bias to the left, or by use of the 'trim' it is possible to bias the stick neutral so that the servo is offset and the control stick then left in the centre of its travel. Trims control about 10% of the total travel of the servo in most systems.

17

Next item on the transmitter is the aerial. This is a comparatively delicate item; although it feels fairly robust it is easily damaged by mishandling and then can fail in its task of radiating the signals developed in the transmitter. Some systems feature aerials which retract fully into the case, some feature case-top mounted aerials. There is no best way.

Some systems are fitted with a meter in the case front. Two types of meter are commonly used. In the first instance the meter quite simply indicates the state of the batteries in the transmitter, usually having a green 'Go' sector and a red 'No Go' sector. The second type is an actual output meter. In neither case will the meter indicate anything until the transmitter is switched on, when the simple

The standard system of indicating transmitter operating frequency for 35 mHz band is an orange pennant with channel number. Battery state meter can also be seen on transmitter fascia.

battery-state meter should go straight into the green, but the output, or RF, meter will only register a small amount of output until the aerial is extended or touched; when the aerial is extended the meter should go straight up into the green. With either type of meter, the manufacturer's instructions concerning its indications should be strictly adhered to, and a habit made of checking the reading before each and every flight. Finally our basic transmitter will have either a hatch for access to the battery box, or, if fitted with rechargeable batteries, a socket for the charger to be plugged into. With every passing year the simple basic system seems to become more of a rarity, as manufacturers vie with one another to make their basic systems more attractive to would-be purchasers. First and probably most useful of the additions made for the novice is the provision of a pupil/teacher or "buddy-box" facility. With this system, it is possible to plug together two transmitters of the same make and type so that the need to pass the transmitter back and forth between pupil and teacher is removed. Different manufacturers' equipment cannot be intermixed to perform this function, even if they both have the same type of plug and socket. There are no national or international standards for R/C equipment yet.

More exotic (and expensive) equipment may have facilities for reversing control throw direction, reducing total control throw, or mixing together different functions, all from the transmitter. From many points of view these 'expert' systems are best left to the experts – content yourself with a basic system until all the controls are fully understood and mastered, and you have learnt to fly your model.

Modern R/C systems all work on a system of digitally encoded pulses being transmitted in a specific order. Each pulse is aligned to one axis of each control stick and in turn to one control surface on the model. The control stick moves a potentiometer (variable resistor) which alters the length of its pulse. This difference in pulse length is detected at the servo, which moves to obey the signal, thus moving the control surface in sympathy with the control stick. Information can be transmitted in either Amplitude Modulated (AM) or Frequency Modulated (FM) form. FM systems are currently more popular as they are less prone to interference, but there is no operational difference in using AM or FM equipment.

Receiver

It is the receiver's task to detect, purify, amplify and decode the signal sent by the transmitter. Detection of radio signals is a comparatively simple task, but the sky around us is jam-packed with signals of every frequency and strength available, and if the R/C receiver is to detect and respond to only that signal sent by its matching R/C transmitter, it must be treated with the respect due to a piece of precision electronic equipment. Modern R/C receivers are little bigger than a matchbox, but the integrated circuits used contain hundreds of resistors, transistors and capacitors.

Once the signal is detected, it is filtered to remove unwanted background noise and any other interference that may be present, then amplified to a level that can be dealt with by the decoder section. The decoder sorts out the chain of pulses and sends them to their respective servos. How this is done is quite simple. Firstly, the decoder 'looks for' a pulse which is longer than all of the others (known as the synchronisation pulse). A typical decoding sequence might be as follows – Sync pulse, Pulse 1 – Elevator, Pulse 2 – Rudder, Pulse 3 – Ailerons, Pulse 4 – Throttle, Sync pulse, Pulse 1 – etc., always in the same order and always taking the same length of time for each 'frame' of information, normally around 25-30 milliseconds (1/1000 sec). Once the decoder has recognised a long 'sync' pulse, it routes the pulses that follow, strictly in order, to the elevator servo, rudder servo etc. until it recognises the next 'sync' pulse whereupon it restarts the sorting cycle.

The receiver will have a flexible wire aerial fixed to it, anything up to 1 metre in length. In most receivers this aerial is a tuned length and must not be cut down or coiled up. Either of these actions can result in the range of the receiver being dramatically decreased. Nor is it advisable to add to the aerial length, which will have a similar disastrous effect on performance. Receiver aerials must be kept away from sources of potential interference, and the electric motors in the servos are a potential source of 'noise'. It is easiest to route the aerial out of the aircraft fuselage as near as practically possible to the receiver and take it to somewhere convenient, e.g. the top of the fin.

Servos

Servos are the muscles of the R/C system, and inside the tiny moulded plastic case there is an amplifier, electric motor and reduction gear box. The amplifier electronically detects the position that the drive arm or disc has adopted via a device called a feedback potentiometer, compares this position with the position described by the pulse sent from the transmitter/receiver combination, and if necessary switches on the electric motor one way or the other to drive the output arm to the position commanded by the transmitter operator. Very fine differences in position can be detected, less than 1° of rotation in most instances, and the servo **can therefore very faithfully follow the smallest of control stick movements.**

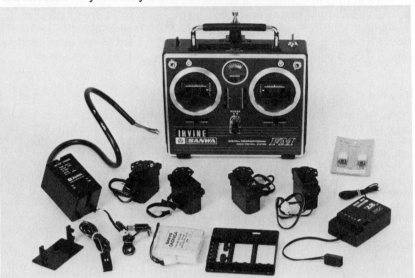

Modern R/C systems are styled like expensive Hi-Fi equipment. This outfit includes transmitter, receiver, 4 servos, receiver battery pack and switch, mains charger, crystals and accessories.

For normal sport flying, speed of travel of the servos is not vitally important; it is more important that the servo is robust and reliable. It is not necessary to seek out the smallest, lightest or fastest. Servos are usually the weakest link in the system, because the electro-mechanical elements, i.e. the electric motor and feedback potentiometer, are susceptible to vibration and do wear out after prolonged use. If the servo is excessively loaded by poor linkages the motor and amplifier can be damaged. Treat servos with care, and many hours of use can be expected from them.

With basic, no-frills, standard equipment, manufacturers normally supply two types of servo, one for clockwise rotation and one for anti-clockwise rotation. Different coloured labels usually distinguish the two. This may at first reading seem irrelevant, for surely the servo rotates which ever way the stick on the transmitter is moved? Yes and no; if for example the servo turns clockwise when the stick is pushed to the right, it may result in the control surface moving the wrong way. It may not be possible to rearrange linkages to compensate for this, so the only alternative is to use a servo which rotates anti-clockwise with RH stick pressure.

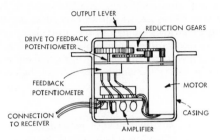

Fig. 9

Batteries

Some reference has already been made to the possibility of R/C equipment being fitted with either rechargeable or 'dry' batteries. As the Nickel Cadmium (nicad) cells typically used for rechargeable systems are initially an expensive item, if included in the system along with the necessary charger and connecting leads, the cost of the system will be increased by a considerable amount. It would not be unreasonable to expect a 20% increase in price for R/C systems in the basic four-function bracket if made rechargeable. Many manufacturers try to make the system more attractively priced by providing the customer with the option of both types of battery, such systems being convertible at a later stage if a dry battery system is purchased in the first instance. In broad terms, most sets use eight cells of the pencell (HP7) size in the transmitter and four of the same type for receiver and servo power.

Two-function systems are able to operate on ordinary high power-type dry batteries, but for three- or four-function operation alkaline type cells are recommended. Dry batteries will give a life of up to four hours in most transmitters, but some four-function transmitters have greater output power and thus need the greater capacity of the alkaline cell. Alkaline cells are not cheap, in fact after buying three sets of alkaline cells for a receiver, you will have spent the equivalent of a set of rechargeable nicads, which will be capable of being recharged well over 100 times. If at all possible, do consider paying the extra price of rechargeable batteries, as they will pay for themselves time and time again in saving on the purchase of dry batteries. Nicads do have other advantages, since they are capable of producing much higher current flow than ordinary dry cells, an advantage if a servo is overloaded, for the dry batteries will very quickly give up, whilst the nicad keeps on giving out power. Most nicad cells are also capable to being fast-charged via a suitable charger from a 12-volt car battery. This can be a great help if mains power is not available, e.g. out at a flying site. A mere 20-30 minutes is all that is necessary to charge fast-charge cells to their full capacity.

Care and maintenance

So much for the equipment, what of its care and maintenance? By and large, R/C equipment is maintenance-free – keeping the batteries fresh or fully charged and keeping the components clean is all that is necessary. Cleanliness is important and cleaning is about the only servicing the unqualified can attempt. Plugs and sockets are frequent sources of trouble if allowed to get dirty. An aerosol of electrical switch cleaner is ideal for sluicing out the inside of sockets, removing grime from plugs, and traces of oil and grease from crystal pins. The lubricant this material contains does no harm either, particularly to the contacts of the receiver on/off switch.

Do keep the transmitter aerial clean, as a dirty aerial becomes difficult to collapse and the extra pressure needed can be sufficient to damage one of the sections. Receiver aerials bear examination from time to time, particularly at the exit point from the receiver case, and the exit and anchorage points on the model. If the aerial has become chafed, return the receiver to the service agent for a new aerial and re-tuning.

Constant removal and replacement of plugs and sockets can cause problems, wires become fractured and the 'spring' goes out of contacts in sockets. A periodic return of the whole system to the service agent will ensure that these items are checked properly.

Often the smart decoration on transmitter cases is screen printed with attractive colour designs which are easily marred by constant contact with oily hands. A wipe over with detergent or, in extreme cases, methylated spirits, usually brings the case up to as-new condition.

Needless to say, never ever meddle with internal adjustment to the transmitter. Not only can such activity degrade the performance of your equipment, it can cause the transmitter to interfere with other people's equipment, even those on other spot frequencies. Leave such adjustment to qualified technicians with the appropriate equipment. Finally, no licence is needed for operation of R/C equipment in the UK on the frequencies laid out in the table.

CHAPTER THREE

Engines

Most instruction manuals supplied with engines deal with the problems of starting and running in a cursory fashion, glossing over the problem which causes more trouble to the novice than any other aspect of R/C modelling. To the expert there are very few motors that will seem difficult to start and adjust, but it must be said right at the outset that starting engines is a knack. Many novices spend many fruitless hours endeavouring to start motors backwards, with incorrect fuel, no booster battery on the glowplug, or with the fuel mixture control needle valve closed.

A number of logical steps must be gone through prior to actually running a new engine, and they are:
(a) Mounting the engine on a suitable test stand.
(b) Arranging a supply of suitable fuel.
(c) 'Loading' the engine with a propeller (which will also aid cooling).
(d) Connecting a booster battery for the glowplug.
(e) Providing a starter for engines of .30cu.in. upwards.

There are several 'test stands' on the market. However, by far the most practical method is to purchase a suitable cast-aluminium mount and attach this

Initial running-in of engines should ideally be conducted on a test stand. This 4 stroke motor seen accompanied by electric starter and glow plug booster battery pack.

23

with appropriate screws to a substantial box, old chest of drawers or even a large baulk of timber which can in turn be fixed in a vice or weighted down in a suitable place. Make certain that the mounting is not distorted when fixing it, as when the engine is bolted to the mount it could in turn be distorted. If all else fails the engine can be run in whilst bolted into its eventual resting place, the model. It is far easier, however, to start an engine if it is firmly bolted down to a rigid test stand, as then the operator's entire effort can be concentrated on starting the engine without being concerned with trying to keep it in one place.

Use the largest size steel bolts that will fit the mounting lugs on the engine; normally these are drilled for 6BA (3.0mm) for up to 0.40cu.in. engine and 4BA (4.0mm) for 0.60cu.in. engines. Some 0.40cu.in. engines will take 4BA bolts. Always use four bolts evenly tightened down; failure to follow this practice can result in the engine crankcase becoming distorted.

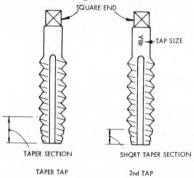

Fig. 10

Propellers

Choose a suitable propeller for running-in the engine. This is not as simple as many would suppose. Most modern engines are made to very high tolerances and very fine finishes on all parts. Considerable development on the metallurgy of model engines has been carried out over the past few years and current thinking is that engines should be run in at their expected ultimate operating r.p.m. in order that all the running parts should bed down properly. During the running-in period the engine should not be loaded down with an over-large propeller or allowed to become overheated, and should be really well lubricated. To satisfy all of these conditions a smaller than normal propeller is fitted and the engine is allowed to run with a very rich mixture setting, that is far more fuel/oil mixture than the optimum is fed to the engine. Some sizes are given below for various engine capacities. Only use wooden or glass-fibre or glass and nylon propellers.

.10cu.in. 7 × 4
.15cu.in. 7 × 4
.20cu.in. 7 × 6
.25cu.in. 8 × 6
.40cu.in. 9 × 6
.60cu.in. 10 × 6

Carefully drill or ream the centre hole of the propeller to suit the engine. Make quite certain that the hole is square to the face of the propeller boss. Carefully

24

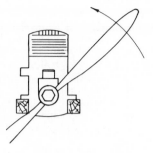

Fig. 11

remove the very sharp edges of the propeller with glasspaper or emery cloth. Bolt the propeller into place using the correct size spanner on the nut. Do not omit the washer. Position the propeller so that it is in the 'ten past seven' position with the engine on compression.

Booster Battery

A 1½ or 2 volt battery will be needed for heating the glowplug element during starting. Dry batteries can be used, but while they are initially cheaper, they only last a very short while. A rechargeable battery is a must for the serious modeller. Many 2 volt 'wet' lead acid cells are available but these are tending to be replaced with sealed 'gel cells' which are totally unspillable and never require topping-up with distilled water.

Nickel cadmium cells of 1.2v are also widely used. A system currently becoming widespread is the 'power panel', which operates from a rechargeable 12 volt battery and provides a metered output at 1.5 and 2 volts, also 6 volts for an electric fuel pump plus sockets for a 12v starter. Some even have provision for fast charging Tx and Rx nicad batteries. Whatever system is chosen a meter indicating glowplug current consumption is a must if any degree of reliability is to be achieved with engine starting.

In conjunction with the battery a means of connecting to the plug is needed. There are many suitable devices on the market, so choose the most robust looking that you are able to find. It is almost inevitable that it will be trodden on during its life at least once!

Starting up

Fix a piece of wire to the throttle lever and either lock the throttle open or arrange a friction clamp so that the throttle can be manipulated during running-in. Fill up the fuel tank with suitable fuel, either a 'straight' mix of 4:1 or 3:1 methanol/castor oil or a similar mix with up to five per cent nitro-methane. Nitro-methane improves starting and idling with an increase in power as a bonus. Some would view the benefits of the addition of nitro-methane in the reverse order, more power with the bonus of easier starting and better idle!

25

Block off one of the filling vents, which prevents the siphoning out of fuel whilst the engine is running.

Remove the glowplug from the engine and check that it provides a good red glow when in circuit. Note the meter reading; most plugs will draw around 3 amps. Squirt a small amount of fuel on to the plug and note what happens to the current consumption. As the element is cooled by the excess fuel its resistance is decreased and the current flow rises. As the fuel burns off the element the current flow decreases. By noting the 'flooded' reading and the 'dry' reading it should be possible to judge from the meter whether the engine has the correct amount of fuel within the cylinder for starting. Too much and it will not run, dry and it will not run. Whilst the plug is out, squirt a few drops of fuel through the plug hole on to the piston, then replace the plug.

Two types of glow-plug are generally in use non-shielded (left), shielded (right). The latter type can be a help in cases where engines are reluctant to idle.

Open the fuel mixture control needle valve in accordance with the manufacturer's instructions, usually about three turns anti-clockwise. Reconnect the glowplug and prepare to flick over the engine.

Almost the whole secret of starting an engine by hand is in the knack of that sharp flick. A really good hard flick incorporates the whole of the arm and a good part of the rest of the body from the waist up. A half-hearted flick is likely to result in a back-fire and a sharp rap on the knuckles, or a kick-back and a bruise on the flicking finger. Put your heart and soul into the flick and the result should be a burst of life from the engine.

Should the engine fail to fire, try again half-a-dozen times but no more. If the engine still fails to fire, prime again through the exhaust port. Grip the propeller firmly and turn over compression. The engine should 'kick' when turned over, the kick being in fact the engine firing. If the engine kicks, flick it over hard and a burst of running should result. If the engine does not kick inspect the meter monitoring the current to the glowplug. If the current flow indicates that the engine is too 'wet', disconnect the plug and flick over several times to clear excess fuel. Also close the fuel control needle valve by a turn or so. If you should suspect that the

Typical Rossi engine components

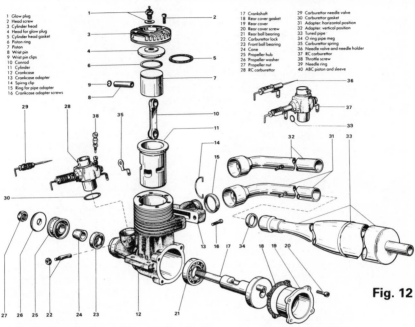

1 Glow plug
2 Head screw
3 Cylinder head
4 Head for glow plug
5 Cylinder head gasket
6 Piston ring
7 Piston
8 Wrist pin
9 Wrist pin clips
10 Conrod
11 Cylinder
12 Crankcase
13 Crankcase adapter
14 Spring clip
15 Ring for pipe adapter
16 Crankcase adapter screws

17 Crankshaft
18 Rear cover gasket
19 Rear cover
20 Rear cover screw
21 Rear ball bearing
22 Carburettor lock
23 Front ball bearing
24 Cone
25 Propeller hub
26 Propeller washer
27 Propeller nut
28 RC carburettor

29 Carburettor needle valve
30 Carburettor gasket
31 Adapter: horizontal position
32 Adapter: vertical position
33 Tuned pipe
34 O ring pipe meg
35 Carburettor spring
36 Needle valve and needle holder
37 RC carburettor
38 Throttle screw
39 Needle ring
40 ABC piston and sleeve

Fig. 12

engine is dry a further prime can be given.

When the engine has been run a few times it becomes relatively easy to judge just how much of a prime it requires.

Electric Starters

Many novice modellers feel that direct personal contact with the propeller of their engine is fraught with too many hazards, even when using a finger protection device, and prefer to use one of the many electric starters available.

An electric starter is not a magic wand, and it will only start the engine if all the conditions within the engine are right. Otherwise it can only make the engine more difficult to start, as the sustained high speed rotation imparted by an electric starter will draw fuel continuously into the engine. If the mixture control needle is incorrectly adjusted the result will be simply to pump fuel from the fuel tank through the engine and out of the exhaust port without it once catching fire. Before attempting to start the engine fit a suitable spinner to the engine. Although most starters will turn over the engine without a spinner being fitted, slipping contact can damage the front faces of the propeller blades and thus make the propeller unsafe. Ideally, use a metal spinner, as this will last for many times as long as the plastic type if repeated use of a starter is envisaged.

27

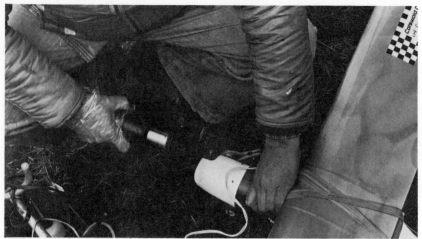

Electric starter about to be used. Note that the glow-plug lead is held safely away from the propeller.

Proceed to start the engine with an electric starter in exactly the same way as described for hand starting, including turning the engine over with the glowplug connected and feeling for the kick. When conditions are felt to be right, place the starter against the spinner and 'pulse' the on/off switch; avoid long periods of application. When the starter is withdrawn after a short burst an apparently dead engine will be found to be running. If the engine does not start immediately check that conditions are right. Should the engine be felt to be flooded, i.e. over-full with fuel, disconnect the plug and close the fuel mixture control valve then spin the engine with the starter until it is clear. In extreme cases it will be necessary to remove the glowplug to pump out excess fuel. Exercise great caution if the engine cannot be turned over by hand, do not apply the starter, damage to the engine can result if the engine is turned over against an excess of virtually incompressible fuel. When the engine is clear recommence the starting procedure but with the needle valve opened a lesser number of turns.

When the engine finally starts running, adjustments should be carried out in the same way as for hand-started motors.

Fuel mixture strength

Even assuming that the engine starts after the first few flicks it is unlikely that the fuel mixture strength will be correct. Either the engine will speed right up then stop or it will run unsteadily with a very smoky exhaust at comparatively slow speed. The former symptoms indicate too little fuel and the fuel mixture control needle must be opened. The latter is the situation that is to be preferred. If the engine continues to run, turn the fuel mixture control needle valve clockwise a small amount and note the increase in running speed. Continue to rotate the needle valve and a point will be reached where the engine appears to switch from a comparatively rough 'four-stroke' to a fast even 'two-stroke'. Adjust the needle valve so that the engine stays at the fast four-stroke setting. At this setting there is plenty of cooling and lubricating fuel passing through the engine and it is running fast enough actually to bed down the parts without unduly stressing anything.

As soon as the engine will run without the booster battery connected to the glowplug the battery should be disconnected. This action will improve the life of the glowplug and also the booster battery.

Running-in

Do not allow the engine to run for more than approximately five minutes for its first run. When the engine has been stopped, ideally by closing down the throttle then pinching off the fuel supply, check over all the mounting screws. Allow the engine to cool for a few minutes, then restart. After about half-an-hour of running the engine it should be possible to close down the fuel mixture control to a 'leaner' mixture strength and change to the flying propeller. If the engine shows any sign of slowing down or overheating, immediately open the needle valve to richen the mixture. If this action is necessary continue to run the engine on the rich four-stroke setting on the running-in propeller for a further few tankfuls before attempting to run the engine flat out.

During this running-in period it is quite probable that the throttle response of the engine will be disappointing. Firstly the engine will not respond to the throttle properly until the high-speed mixture strength setting is correct. Secondly the engine will not idle properly until it is fully run in. Running-in is complete when the engine will run at its maximum speed without slowing down or overheating on the correct-sized flying propeller for a full tank of fuel.

In spite of engine manufacturers' advice to the contrary, running-in of engines in the model and flying around does not seem to achieve the same results as can be achieved by careful bench running. If a mistake has been made in the setting of the needle valve and the fuel mixture is set too weak, the anxiety of flying what is perhaps a new model plus the almost inevitable confusing sounds of other modellers' engines can prevent the signs of distress and imminent engine seizure from being heard until too late.

Properly run in, a modern two-stroke model engine is capable of giving several years of good service with virtually nil maintenance.

Mounting the engine

Fitting engines to metal mounts is strongly recommended. This is not difficult, and the following description should help the non-engineer to complete the task satisfactorily. Purchase the appropriate size drill selected from the table (fig 14) for the purpose in hand. It is not necessary to purchase High Speed Steel drills, the cheaper carbon steel alternative will do for this job if obtainable. At the same time obtain a suitable size tap. Once again, carbon steel is suitable. Ask for a 'second' tap. This does not refer to quality, but to type; a second tap incorporates a slight taper which eases the entry into the metal being cut. Something to turn the tap will be needed, a 'Tee' pattern tap wrench being the most suitable. Finally a large

'Obo' nail will perform as a centre punch, but if funds will stretch of course purchase a centre punch.

Proceed to mount the engine as follows:

1. Clamp the engine to the mount. In some cases it will be possible to turn the engine upside down on the mount to ease the marking.
2. Mark through the mounting holes in the engine on to the mount. Use a drill held in the fingers that will just clear the holes. Twirl it round until a mark has been made on the mount. Alternatively use a pencil or the Obo nail.
3. Remove the engine from the mount and place the mount on a solid support. Carefully punch the marks indicating where the holes are to be drilled using the Obo nail and a hammer.
4. Using the appropriate tapping size drill, bore right through the mount. It is a good idea to start all the holes then compare with the engine before drilling all the way through. If any error is present it can be corrected by drifting the centre of the hole to one side with the punch and a hammer.
5. Place some lubricant, paraffin or a very thin oil such as '3-in-One', on the tap and applying light pressure, turn the tap clockwise in the hole. It should start to bite and draw itself into the hole. Continue to cut the thread, advancing the tap half a turn forward and then a quarter turn back to break up the chips of metal.
6. When the tap has cut right through the mount, remove it and clean out all traces of metal swarf from the threads. Clean the tap and lightly oil it to prevent corrosion before it is next needed.

Should it prove necessary to adjust the thrust-line of the engine, do not pack out the engine from the mount or all the effort made will be wasted. Pack out the engine mount from the bulkhead to effect alterations necessary.

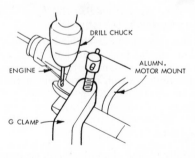

Fig. 13

Fuel tank installation

Many problems relating to poor engine performance, bad idling and erratic starting can be traced to poor or faulty tank installation. The majority of fuel tanks used are of the polythene bottle 'clunk' type. They are arranged in such a way that the engine is able to draw in fuel in whatever attitude the aircraft assumes, achieved by using a flexible pick-up pipe fitted with a weight so that the end of the pipe will always 'clunk' to the 'bottom' of the tank, thus following the fuel. Two additional tubes are usually fitted to fill the tank and simultaneously allow the air to vent from the tank. Complete assembly instructions are usually supplied with such tanks, but several additional points are worth mentioning.

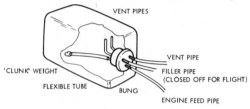

Fig. 14

Before installing the tank, check that the flexible pick-up pipe is not too long, so that the 'clunk' weight can easily drop to the bottom of the tank whichever way the tank is turned. It is worth checking that the pipe is actually flexible enough to allow this to happen. If not, replace it. Also check that there are no sharp 'burrs' on the ends of the pipe that can cut holes in the flexible tubing, either inside or outside the tank. Finally check that moulding 'flash' on the the sealing parts of the tank does not prevent the tank from sealing, thus enabling fuel to leak into the tank compartment.

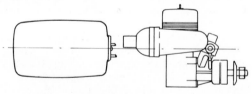

ENGINE AND FUEL TANK ALIGNMENT   **Fig. 15**

When satisfied that the tank is properly assembled and leak-proof, instal it into the model. The centre line of the tank should be level with the jet of the carburettor. If the tank is mounted too high, the fuel will siphon into the engine, if too low, the fuel mixture strength can vary to an unacceptable degree during the course of the engine run.

Pressure Systems

Many modellers compensate for less than ideal tank positioning by employing pressurisation of the fuel system.

Many silencers fitted to engines incorporate pressure bleed nipples and if they do not it is a simple matter to fit one by tapping an appropriate-sized thread into the silencer and screwing in a suitable nipple, easily obtainable from your model shop. If the specialist item is not available, the fuel connection from a carburettor is an excellent substitute. Identify the threads by screwing the nipple you intend to use into a selection of nuts of known thread size and then tap a thread into the silencer. The position is not critical, simply choose a position which suits the tank and engine installation and which will need the minimum of flexible tube connection.

Fuel Filter

No full-size aircraft, car or motorbike, or even a lawnmower, would ever be designed to operate without a fuel filter. How so many modellers expect to enjoy trouble-free operation of their engines without fitting such a simple device is a continuing mystery. Carburettor jets on model-size engines are far smaller than those employed on full-size engines which, coupled with the comparatively filthy conditions that they are filled up with fuel and operated in, make it often surprising that they are able to run at all. Many filters are available, most are inexpensive. As a very minimum at least one filter should be fitted between fuel tank and engine. It is advisable to filter the fuel from the bulk supply into the tank on the model as well. Remember, if using a reversible fuel pump to drain a fuel tank, to reconnect the filters so that particles of rubbish are not pumped back into **the bulk fuel but remain trapped in the filter.**

Some form of fuel filtering is essential – author favours fitting a filter in the fuel line.

After every flying session clean out the filter. Some are designed to be dismantled for cleaning, but others can only be cleaned by back-flushing with fuel. Whichever type of filter is used, clean it.

A simple additional precaution which can save an engine cut on take-off and the resulting heavy landing or crash is the insertion into the dismantleable filter of a tiny piece of plastic foam of the type used for cushion mattresses etc. Make sure, however, that the wire mesh filter is left in the filter housing between the foam and the engine, or the minute piece of foam can be drawm into the fuel line and thus shut off the fuel supply. This additional foam filter is capable of filtering out the minutest of particles and costs virtually nothing in time or expertise, so replace for each flying session.

Carburettor adjustments

Once the engine is running, open the throttle fully and adjust the needle valve for maximum r.p.m., then open up the needle valve approximately one quarter of a turn. Tip the nose of the model up to at least 45° and check that the engine does not falter. If it does, open the needle valve further until it will still run smoothly without loss of r.p.m. in a steep climb attitude.

Once the carburettor is adjusted for full speed running the following checks should be made:

1. Close the throttle and see if the engine dies down to a smooth idle. If it cuts out straight away it may be that the idle speed you are expecting is too low. Try idling the engine at a slightly higher speed. If a tachometer is available aim for an idle speed of 2500-3000r.p.m. If the engine will idle comfortably with the transmitter throttle control stick not pushed fully down, then adjust the throttle stop screw accordingly. You must now check that full and free travel of the throttle linkage is still available.
2. If the engine will not idle at all observe carefully the manner in which it stops. It will almost certainly speed up slightly then stop, giving the appearance of being starved of fuel as if the needle valve had been closed too much, or it will run smoothly and smokily then splutter to a stop.

In the first case the idle mixture is too weak, that is the ratio of fuel to air is biased towards too much air and not enough fuel. What is to be done to rectify the situation? Quite simply, adjust the carburettor! It sounds so simple but provides many modellers with a major dilemma. There are as many types of carburettor available as there are engines, and while all have broad similarities, details vary from make to make. Generally speaking, the fuel/air ratio requirements from peak speed to idle will vary – a weaker mixture will be needed for idle, that is more air and less fuel. A gradual progression from the enriched mixture to the weaker idle mixture is the ideal and some modern R/C carburettors are designed to provide just this. Most offer a compromise on the middle ranges and correct mixture adjustment for full speed and idle. Various systems are used, viz:

1. Fixed air bleed (to weaken mixture at idle), adjustable full speed (main needle valve).
2. Adjustable air bleed, by far the most common. Adjustable full speed (main needle valve).

3. Automatic mixture control by controlling the idle speed fuel flow rather than the air flow. Adjustment facility for full and low speed.

Early carburettors were largely of type 1, but it soon became apparent that a simple means of adjustment to the air bleed would expand the idle speed operating range of the carburettor. Automatic mixture control carburettors attempt to vary the fuel/air ratio throughout the range, using various mechanical means. The exact method of achieving this state of affairs is largely irrelevant to the adjustment of the carburettors and it suffices to say that the individual carburettors regulate the mid-range fuel mixture with differing degrees of efficiency, depending on design. It should be said, though, that a carburettor badly adjusted at either end of the full range will not perform as its designer intended in the mid-ranges.

Several carburettors now feature mid-range adjustment facilities which just means one more thing to go wrong for many modellers!

Having briefly outlined the types of carburettor available, and said what they are intended to do, an approach to adjusting any carburettor is needed. The following principles broadly apply:

1. With the aid of the engine instruction leaflet identify the various adjustment points of the carburettor. Locate the throttle barrel stop, the air bleed adjusting screw, the mid-range adjusting screw (if fitted) and the idle mixture fuel control (if fitted).

2. Assuming that the engine is running at idle, however roughly, rapidly open the throttle and observe what happens. There are two likely consequences: (a) the engine will start to pick up then just die away (mixture too weak) or (b) it will splutter and smoke and slowly clear itself, then pick up (mixture too rich).

3. (a) For Air Bleed Carburettors. If the mixture is too weak screw in (¼-turn at a time) the air bleed screw, thus shutting off the air supply and richening the mixture. Repeat step two again, observing the result. Repeat the adjustment until the engine picks up cleanly without hesitation to full r.p.m.

It may be necessary to re-adjust the throttle stop screw when this adjustment has been completed. Check now for full throttle operation and re-adjust as necessary. Should the mixture be too rich, unscrew the air bleed screw, thus allowing more air to flow. It should be quite possible to adjust an air bleed screw-type carburettor for good clean throttle response with no fear of an engine cut however quickly the throttle is opened and closed. It may be noticed that the middle range of speed is less than a 'clean' two-stroke but this is a limitation of this simple carburettor and cannot be 'adjusted away'.

3. (b) With Automatic Mixture Control Carburettors without air bleed, usually turning the idle mixture needle clockwise weakens the mixture and vice versa. Refer to the engine instructions if in doubt, or poke and hope – after all, if you only turn the screw ¼-turn clockwise and it's wrong, you can always turn it back again. The important thing to remember is that smoking and spluttering on opening up indicates rich and a 'clean death' is weak.

It is worth mentioning that adjustments on carburettors such as the **Kavan** and **Perry** need only be moved the minutest fraction to alter the idle mixture strength. The **Perry** system is friction-locked whilst the **Kavan** has a locking screw – both are adjusted in a similar way, that is the whole barrel containing the fuel metering device is rotated. Automatic mixture control carburettors such as the OS Irvine,

Webra TN and Dynamix are all fitted with an adjustable needle valve for idle mixture control.

4. Automatic Mixture Control with Air Bleed. A typical example of this type of carburettor is fitted to the Meteor 40 and 60 engines. This carburettor has an adjustable air bleed for final idle mixture strength compensation after mid-range adjustment has been completed. After adjusting the full throttle setting, assuming the engine will keep running at low speed, adjust the mid-range needle for smooth running at the half-throttle position. Rough four-stroking or four-stroking breaking into high speed two-stroking occasionally indicates too rich a mid-range setting and the control needle should be adjusted accordingly, clockwise for weaker and vice versa. Proceed to adjust the air bleed as for any other air bleed-type carburettor after the mid and top end are satisfactory.

Should it prove difficult to keep the engine running without first adjusting the air bleed, a preliminary attempt should be made at this adjustment and returned to when the other ranges are satisfactory. In a nutshell the principles of adjusting a carburettor are:

(a) Adjust the full-speed mixture first with the throttle wide open.

(b) Establish whether or not the idle mixture is weak or rich and adjust.

(c) Adjust idle mixture until a clean pick-up and reliable idle are obtained.

(d) Do not attempt to adjust the idle mixture with the main needle.

(e) Always check that any adjustment made to the throttle stop does not foul up the linkage to the servo.

It may be found that however carefully the mixture controls on the carburettor are adjusted, the engine will not idle properly or pick up cleanly. It may be that the glowplug being used is not able to sustain a good enough glow at low r.p.m. or cope with the sudden increase in fuel flow when the throttle is opened. The sudden increase of fuel flow which occurs when the 'tap' is opened is akin to just that – a deluge of cooling fuel on the plug element which causes the "fire" to go out. Fitting an "idle-bar" type plug usually helps to cure this problem.

Modern kit, the Piper Tomahawk by Hegi, includes g.r.p. fuselage, vac-formed cowl, veneered foam wings, etc. Span is 1817mm (71¹/₂ in.)

Traditional all-balsa construction is shown in this typical trainer. Sheeted leading edge, capped ribs and double t.e. are all anti-warp measures.

Construction

There seem to be several distinctly different combinations of circumstances that combine to tempt people to take up R/C modelling. Many are exposed to it almost by accident, either seeing a demonstration at a fete, or maybe just coming across a group of modellers during a country walk or drive. This first exposure quite frequently dictates the type of model that the individual finally chooses to build and fly. Those who are keen walkers might well first encounter slope-soarers at sites such as Ivinghoe Beacon, the well-known Chiltern beauty spot. There is a certain amount in common between slope-soaring enthusiasts and hill walkers – both appreciate the copious fresh air, and the peace and wide-ranging view of the high, open places suitable for slope-soaring. Visitors to fetes will frequently see only power flying, either conventional fixed-wing aircraft or rotary-wing types (helicopters). Engines might not appeal, and hill-climbing might not either, so the obvious choice is a thermal-soaring glider which can be flown in silence from a flat field site.

Of course R/C modelling, and indeed many other types of modelling, can be an expensive activity, and it is common for many people temporarily to drop their hobby activities during early married life, but later return. These latter types frequently know very clearly what they would like to do, having previously experienced modelling, and, knowing many of the likely pitfalls, are able to approach their new start in the hobby from a positive point of view.

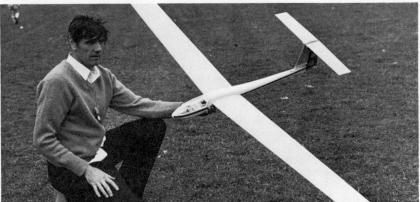

Elegant Almost Ready to Fly (ARTF) Thermal Soaring glider with moulded plastic fuselage and veneered foam core wing. Ivory Gull with designer Dick Edmonds.

Necessary equipment

Whatever aspect of R/C flying appeals, the first step is to list the necessary equipment for the chosen branch. Slope-soaring requires the least capital outlay of all types of flying; R/C equipment, a model, and that is about all. A would-be thermal-soarer needs to add a means of launching his glider to the shopping list, the power flyer an engine and all the ancillaries. Type of model does of course dictate choice of equipment; a basic slope-soarer can give years of pleasure with only two-function R/C equipment, but three- or four-function R/C might be needed if precision aerobatics or scale slope-soarers are felt to be a future aim. If in doubt, buy more functions than you think you are going to use; it is usually possible to buy, say, a four-function transmitter and receiver with only two or three servos, thus cutting the initial capital costs. Additional servos can always be added as more ambitious models are felt necessary. Some manufacturers of R/C equipment are able to offer systems that can be expanded upwards from two functions to seven or eight by plugging in additional modules.

It is not absolutely necessary to buy the R/C equipment at the same time as the model, but progress on the model will be very limited if the equipment is not to hand to check sizes of servos, positions of control connection and so on. It would be wisest to purchase as much of the equipment at one time as possible, or at the very least restrain the impulse to start building too much of the model before the R/C equipment and engine are to hand.

Once having decided upon the type of R/C modelling that is most suitable, and what type of R/C equipment this activity requires, the actual model type must be chosen. At this particular point in the chain of events it is as well to be resigned to the fact that R/C flying is something that everyone has to learn to do. It is not usually impossible for anyone interested to find a level that suits him with regard to speed of model, aerobatic potential and level of flying difficulty, but it is most important not to attempt to pre-empt the learning process and step over intermediate stages. Providing the right attitude is adopted R/C flying is not difficult, but attitude starts at the moment the choice of first model is made.

First model

Not for nothing does every kit manufacturer in the world list R/C trainers in his range; a trainer is absolutely essential. It may not be the sort of model that is ultimately looked for, but a stage that is gone through. Do not on any account be tempted by glossy box-top pictures of sleek aerobatic models or immaculate scale models – they are for the experienced. Trainers exist for both power and glider flying, and generally they are more robust and simply constructed than the more exotic types, always they are designed to be aerodynamically stable with the ability to fly by themselves if let alone. Most trainers use only rudder and elevator control, and of course throttle control for powered models.

Resigned to the fact that a trainer model has to be first on the list, there are three possible avenues of approach to the problem of actually possessing a finished model. Two of these are reasonable, the third, strictly not recommended, is design your own model. Real choice lies between kit-built or plan-built models. If model building is likely to be a major part of the interest in R/C modelling, building from plans has much to recommend it; cost is also lower than kit building.

Building from plans is really more suitable for the returning modeller, that is the person returning to aeromodelling after a lapse, as plan building usually pre-supposes some basic experience of model aircraft building techniques and probably a wider range of tools than would be needed for kit building. Plans can be found for all types of model, both gliders and power models, in the excellent MAP range (see Appendix for details).

Kits

By far the most common introduction to model building is via a kit, and there are kits available to suit every level of building ability and every depth of pocket. In general terms the more comprehensive the contents and the more prefabricating is done, the more expensive the kit will be. If the thought of complicated construction and finishing is too daunting, 'kits' are available that require little more than the installation of R/C equipment and a coat of paint before the model can be flown. There is a strong tendency towards Almost Ready to Fly (ARTF) models recently, many such models using moulded plastic fuselages, either vacuum-formed or glass reinforced plastic (GRP) and solid plastic (polystyrene) foam wings either uncovered or veneered with hardwood veneer 0.5mm thick.

More traditional construction methods such as the balsa frame, either tissue or fabric covered, usually result in models that are lighter for any given size, which in turn produces a better-flying aeroplane. It is also only fair to say that the traditional balsa/ply model may suffer more in the event of a heavy landing or

Foam and Fibreglass ARTF trainer – note the simple, attractive self adhesive decoration on the fuselage.

crash, though this is partly offset by the fact that it will usually be much easier to repair. The choice is really for the individual builder, quick building and a tough, heavier aeroplane, or more involvement in the building stages and potentially better flight performance.

Whichever course is chosen, examine the kit carefully and make quite certain that what is at first sight assumed to be good value is not degraded by the absence of many vital 'extras' such as wheels, fuel tank, linkages etc. Several 'bargains' look less so when these extra items are added to the basic cost. Check also on the engine size requirement, as large engines are expensive, and models for large engines are large! It may be that there is already a basic restriction on the size of the model, imposed by the size of the boot of the family car! Come what may, if a range of engine sizes is specified, err towards the top end of the sizes recommended. After all the model is to have throttle control, which can always be used to reduce the engine power output, but no amount of encouragement will persuade an underpowered model to perform properly.

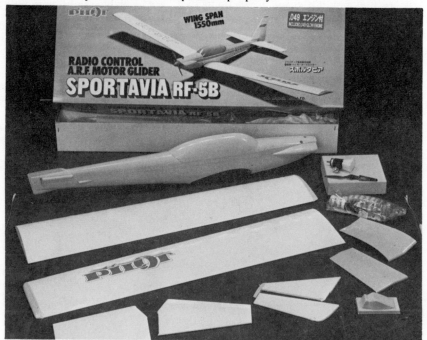

ARTF kit contents in this instance even include the engine and pre-finished wing panels.

Tools and building board

Once the kit is purchased several basic tools will need to be gathered together, plus a flat board and suitable glue for fixing the parts together. Many of the tools required will form part of the average home handyman's tool kit already. The

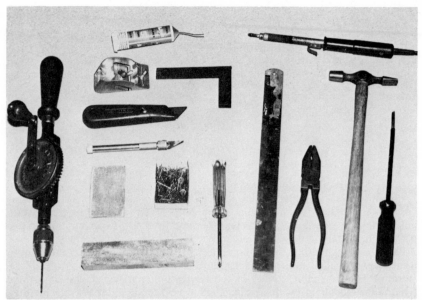

Basic tool kit for R/C aircraft construction. Rectangular blocks have abrasive paper glued to them. A range of sizes and grades is very useful.

following will be needed as a starting point, with desirable extras in brackets following:

1. Stanley knife – for heavy cutting operations.
2. Scalpel – fine work.
3. Steel rule – measuring and cutting guide.
4. Razor plane – carving block parts and producing accurate chamfers.
5. Steel pins – 'safety' glass-headed type or steel dressmaker's type. (Small hammer.)
6. Pliers – engineer combination 6in. (snipe nosed, round nosed, side cutting, end cutting).
7. Drill – wheelbrace type.
8. Twist drills – straight shank jobber's type 1.5-6.5mm in 1mm steps (intermediate size drills).
9. Soldering iron – 30 watt.
10. Screwdrivers – plain and Phillips type.
11. Glasspaper – various grades.
12. Glasspaper blocks.
13. Razor saw (Hacksaw).
14. Square – engineer's or carpenter's.

Final flying performance of the model is determined by the attitude towards building which is adopted from the very outset of construction. Do not be prepared to accept anything other than the very best that you are capable of. Start by finding or making a really good building board; even if the model to be built is

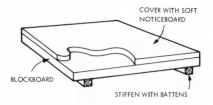

COVER WITH SOFT NOTICEBOARD

BLOCKBOARD

STIFFEN WITH BATTENS

Fig. 16

an ARTF type, it will be necessary at some stage to set things up and check that they are flat, true etc. Blockboard is a very good material to use as a basis and 25mm thickness is suitable, sized to suit the largest model that is likely to be built – very few models have any individual component part larger than 1 metre in length unless they are very large gliders. Bear in mind that most wings are built in two parts, joined later.

Face the blockboard with a softer material that is capable of easily taking pins, which are very widely used in model aircraft construction. A visit to the local DIY store will usually enable a piece of soft board to be purchased, and this can be fixed to the blockboard with an impact adhesive if required, or a lip can be fitted around the blockboard so that the piece of soft board can be easily changed, when its surface becomes marked or damaged.

Once the tools have been gathered together and a building board prepared, the temptation to rush into the exciting part becomes overwhelming.

Glues

Before actually sticking Part A to Part B, consider just what they are to be stuck with. Returnees will probably remember balsa cement, and many newcomers will have come across it. It does certainly have a place, but most R/C modellers do not use it in their workshops, but rather turn to the more modern PVA white glues and epoxy resins. Using PVA woodworking glue will result in a stronger yet more flexible airframe than balsa cement would provide. Being a water-based emulsion, PVA soaks into wood surrounding joints rapidly, producing a very good bond. Used on balsa it has a comparatively rapid setting time, one hour or less, but still sets slowly enough to allow good handling time for jobs such as fixing large areas of sheet balsa which require many pins to hold down. Buy in the largest size you feel able to handle, as bulk buying certainly brings economies.

Epoxy resins are very useful in high-stress areas, but beware of too liberal use of '5-minute' varieties, which set so rapidly that they are often unable to penetrate the wood fibres in the area of the joint thoroughly. Epoxy can be used to advantage for fixing engine bearers, joining wings and other 'tough' jobs.

Most recent of the glues to find its place in the R/C modeller's workshop is cyanoacrylate. Frequently known as 'Instant Glue' or 'Wonder Glue', this adhesive cures so rapidly as to be at times alarming, for coupled with rapid bonding, in the right circumstances it is very strong. There are now many types of "cyano" available, some suited to bonding porous materials, others of lower viscosity better suited to non-porous materials such as metals and some plastics. This glue is very expensive and though some modellers have built entire models using it, super-accuracy in joint preparation is essential if large amounts of glue are not going to be wasted soaked up in joint filler material. Don't be without it, however, now that it is available – benefit from its usefulness as an instant repair agent for pieces of wood accidentally split during building or on-the-field repairs.

Most common adhesives used for model building are l to r twin pack quick-set epoxy, cyanoacrylate (instant glue) and white P.V.A woodworking adhesive.

Cutting out and constructing

Do read instructions and study the plan of your model thoroughly before starting, making sure you are completely familiar with the order of assembly and have identified all parts correctly. Nothing is more frustrating than to find that the long strip of wood cut into two pieces at an earlier stage was in fact the long piece required later, and there were two short pieces in the box! The old axiom 'think twice, cut once' is well worth keeping in mind.

Accuracy is all important; a joint that does not fit properly and is filled with glue will not be as strong as one that is close-fitting. Try to avoid having to spring pieces of wood into place, since this will result in an airframe that has built-in stress than can only relieve itself by distorting the structure. Use a steel straight-edge as a guide to cutting pieces of wood that butt up to one another and always check that

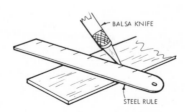

Fig. 17

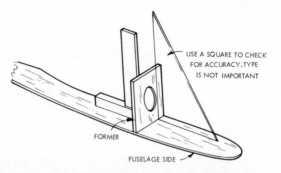

USE A SQUARE TO CHECK
FOR ACCURACY. TYPE
IS NOT IMPORTANT

FORMER

FUSELAGE SIDE

Fig. 18

parts that need to be upright are indeed so by checking with a square.

If the model chosen is to be constructed from a plan, the first task is to estimate the quantity of each size and type of materials needed for construction. Prepare a list allowing for a reasonable quantity of waste, basing the balsa sizes on the following sheet sizes: 36 × 2in., 36 × 3in., 36 × 4in. Strip sizes such as ¼ × ¼ (6.5 × 6.5mm) can be obtained, or they can be cut from suitable thickness sheet using a special balsa strip cutter or a steel straight-edge and Stanley knife. Cutting strip is quite a lot cheaper than buying cut strips.

Once the material is to hand, it will be necessary to transfer the shapes of the individual parts on to the balsa and plywood. Before starting, identify the grain direction specified by the designer, and with this in mind plan out the layout of the parts on the materials. Shapes can be transferred to the material by several means, carbon paper or pricking through with a pin, or if two copies of the plan or a photocopying machine are available, pasting cut-out shapes directly on to the wood with Cow Gum. Use of carbon paper is self-explanatory, pricking through may need explanation; quite simply using a pin, prick through the plan along the outline of the shape to be transferred, close together round the tricky curved portions, just at the ends of straight lengths. Once the whole outline has been pricked out, the pin pricks can be joined together with a pencil line on the wood.

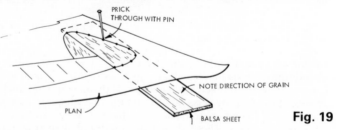

PRICK
THROUGH WITH PIN

NOTE DIRECTION OF GRAIN

PLAN

BALSA SHEET

Fig. 19

Using tracing paper (or kitchen greaseproof) is another possibility, widely used. Although it may seem tedious at the time, it is a very good idea to cut out all the parts before starting, to produce a kit, in other words. Once this part of the job is done, the rest of the building process moves along very quickly.

If an ARTF model has been chosen there will still be some constructional work to be done in most cases. This may take the form of fitting out a GRP fuselage, in which case the foregoing comments about glues can largely be forgotten. To bond the wooden parts to GRP mouldings, a polyester resin must be used, such as those sold for car body repairs. Most hobby shops, however, now stock handy-sized kits of GRP resin and hardener. Follow the suppliers' instructions very carefully when using the resins, as it can be most embarrassing to have nearly slid a bulkhead into place inside a fuselage only to have the resin go 'off', firmly fixing the bulkhead in the wrong place. If resin has to be used, clean the moulding surface thoroughly, using emery cloth or glasspaper to scarify the area of the joint. Don't allow fingers to contact the joint area once it is clean, as the minute amount of grease deposited can spoil the bond.

Another plastic sometimes used, for model glider fuselages in particular, is ABS. This material needs special solvent adhesive normally supplied in the kit. Joining can be made easier by producing a paste of shavings from the plastic dissolved in the adhesive. Normal glues such as PVA and epoxy are not at all suitable for most plastics, and even cyanoacrylate has only limited ability to join them.

The most common means of reducing the time required to build a model is to use polystyrene foam-cored veneer covered wings. Polystyrene foam is a lightweight cellular material with very little structural strength, the necessary strength

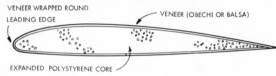

VENEER WRAPPED ROUND LEADING EDGE

VENEER (OBECHI OR BALSA)

EXPANDED POLYSTYRENE CORE

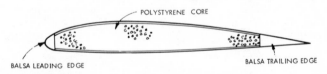

POLYSTYRENE CORE

BALSA LEADING EDGE

BALSA TRAILING EDGE

Fig. 20

being provided by gluing on skins of hardwood veneer. Most aircraft wings are produced in two separate halves, either because of the need to angle them relative to one another for dihedral, or because of their compound thickness and planform taper. It is generally necessary to join the two veneered halves and reinforce the joint. Most foam-covered wings also need leading and trailing edge balsa reinforcements fitted. These should be fitted and shaped as required, before the halves are joined together. Only epoxy or PVA and other non-petroleum based glues should be used, otherwise glues such as impact adhesive will dissolve the foam.

Once the separate wing halves are prepared, they should be butt-jointed at the centre, using either epoxy or PVA glue. The centre join will need to be reinforced, and most kits and plans specify glass-fibre bandage and polyester resin for the reinforcement. Beware, polyester resin also dissolves polystyrene foam, so make certain that the joint area is totally sealed before applying a polyester resin. It is a good idea to mask off the outer portions of the wings to prevent resin spoiling these areas. Surface finish of the centre reinforcement can be considerably improved by stretching polythene sheeting tightly over the resin-soaked glass-fibre until it has cured. If any attempt is made to smooth the area, on no account thin the veneer covering where the reinforcement finishes, as to do this will seriously weaken the wing, probably leading to it breaking under stress.

Above all work carefully and methodically, taking full notice of the instructions, and never be tempted to accept second best. If any part of the structure is not right, whether it be on a kit-built, ARTF or plan-built model, do it again correctly. Extra time spent on constructing a sound and accurate airframe will be adequately repaid in length of service and flying performance.

More traditional open frame structure on this small biplane sports model built from a kit. (Photo G. Palmer)

CHAPTER FIVE

Finishing

Once the basic airframe construction is complete, decisions have again to be made. Probably during the course of construction some firm ideas on colour scheme, marking and lettering etc. will have been formed. There are three possible avenues to follow:
1. The traditional tissue or nylon/silk followed by paint of some description.
2. Heatshrink self-colour, self-adhesive material, either of the plastic film or fabric type.
3. Glass cloth and epoxy resin followed by paint.

Sanding

All three methods have advantages and disadvantages of time, difficulty and durability and these will be outlined in due course. Whatever method is chosen, the finished result will be dependent upon the careful preparation of the model before the first paint is applied. Basic preparation is the same whichever finishing method is used, and is started off by careful, accurate sanding of the wood framework. To some would-be modellers the term accurate sanding may seem a contradiction in terms, but this is not so. Properly used, glasspaper or other types of abrasive paper can produce precise results. The first step is to obtain the correct abrasive papers. Strange as it may seem, the softer the material to be cut the sharper the tool needed, and the more quickly the edge becomes blunted. Glasspaper is a cutting tool, the minute particles of abrasive forming thousands of tiny cutting edges which do become blunted with use. For best results on balsa wood use garnet paper, and you will need several different grades, ranging from a coarse grade for rough shaping to very fine for final finishing. A very good material to use, even if rather expensive, is fabric backed material used for belt sanding machines.

During the course of building a number of models, a range of sanding blocks will be built up, made as occasion demands to fill specific needs. Small pieces of scrap hardwood or balsa can be used for these blocks. Fix the glasspaper to the blocks with impact adhesive.

Using sanding blocks gives much greater control over the cutting rate of the abrasive, particularly where sanding over the junction of two grades of wood with a glue-line in between. Without the abrasive held flat with the block, uneven finger pressure through the sheet will cause unwanted hollows to be produced, and prevent the levelling down of bumps.

When sanding compound curves, such as wing tip blocks, it will be found more

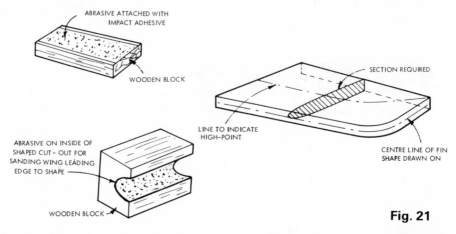

ABRASIVE ATTACHED WITH IMPACT ADHESIVE

WOODEN BLOCK

ABRASIVE ON INSIDE OF SHAPED CUT - OUT FOR SANDING WING LEADING EDGE TO SHAPE

WOODEN BLOCK

SECTION REQUIRED

LINE TO INDICATE HIGH-POINT

CENTRE LINE OF FIN SHAPE DRAWN ON

Fig. 21

convenient to use a piece of abrasive sheet without the support of a sanding block. Internal curves such as wing and tailplane fillets will need circular (cylindrical) sanding sticks. These can be made by wrapping abrasive paper round wooden dowel or even convenient-sized tin cans if size dictates.

Don't be tempted to try and remove too much material with abrasives, not only is it tiring, but it can result in inaccuracies. Instead carve or razor plane away the bulk of the material. If you are unsure of your ability to gauge shapes of leading edges of wings, tailplanes and fins, a simple template cut from card will help to maintain an accurate profile over a long length. In any event, a guide line is an essential for accuracy on thin sheet for such items as rudder, elevators etc. Once the rough sanding is completed, proceed to refine the surface, using successive grades of abrasive. Very fine glasspaper clogs quickly; wet-or-dry carborundum paper clogs less easily on balsa wood and a final rub over with 600 grade

Shaping and finishing tools, the sanding block and razor plane.

wet-or-dry will leave the wood with a satin-smooth surface.

Filling

Sanding will almost certainly have revealed areas that need filling, usually minor gaps in joints and the small but inevitable marks in the soft balsa surface caused by the odd slip, or even placing the part down on a hard object. All of these marks need to be filled before further work can be done. Very bad blemishes can be filled with carefully carved scraps of soft balsa. For the minor marks some form of filler should be used. Any filler should be softer than the surrounding material, otherwise when the attempt to sand it down to a level surface is made, the hard material will be left proud and the surrounding material sanded down still further!

'Plastic wood' is much too hard for filling purposes, but there are a number of proprietary fillers that can be used; one of the cellulose-based plaster-type fillers used for decorating will work well. Epoxy putty specially formulated for the job will do, but as these are usually coloured with a dense pigment, can only be used if a fully coloured finish is to be used, otherwise the dark patches of filler will show through the covering. Many fillers tend to be comparatively heavy, but 'micro-balloons' mixed with resin or clear cellulose dope provides an exception. Micro-balloons (a trade name) looks like a fine white powder, but is in fact comprised of millions of minute hollow glass spheres which, when mixed with the binding agent, sand very easily to a fine surface. It can be used as a complete surface preparation, taking the finishing procedure from bare wood to painting surface with only the single intermediate stage.

Once all surface imperfections are dealt with, a final sand with 600 wet-or-dry should leave the surface requiring only to be dusted thoroughly ready for the next stage.

With an immaculately prepared airframe ready to hand, pause for a while and consider the alternative finishes available

Tissue

One of the oldest traditional covering materials, tissue can form the basis for an excellent finish. Model aircraft tissues are specialised materials and come in quite a variety, from the ultra-light grainless condenser type, through Jap, which has a pronounced grain direction, to the Modelspan and similar rag tissues with omni-directional grain. Only the latter really concern the average radio modeller; there are two grades, light and heavyweight, usually available only in white and there-fore requiring dyeing or colour after application. Both types are used in conjunc-tion with cellulose dope, a varnish-like liquid which both shrinks and airproofs the tissue. Most modellers apply tissue dry, using either paste or dope as an adhesive to fix it to the framework. Once the tissue is fixed in place, it can be initially watershrunk (to which colour can be added), then given several coats of clear dope until the required tautness is achieved. Once shrunk, the doped tissue will take painted finishes well. The resulting covering is strong from a structural point of view, but quite brittle and can be punctured easily, e.g. landing the model on a stubble field can result in a wing resembling a colander.

Nylon

Incredibly tough, nylon is applied in much the same way as tissue, although usually moistened first. Once cut to shape, the piece of nylon is soaked in water, the excess moisture squeezed out and the nylon is then stuck down with dope, stretched into place and held with dressmakers' pins. Using the material moist helps prevent initial coats of dope simply soaking right through the material. It is essential that a very close weave material is used, preferably one that has been specifically selected for covering model aircraft. Once fully doped the material can be painted as required, or if a coloured material was chosen, left in its original colour.

The resulting covering is almost indestructible, and a bad crash frequently results in the nylon covering finishing up as a bag for all the broken pieces! A doped nylon finish is probably the heaviest of all the covering methods available but will probably last longer than any of the others. There is a definite art to applying nylon covering, but the results are worth the effort.

Silk

An intermediate between tissue and nylon, silk is quite strong, but tears in the finished covering will occur far more easily than in nylon. It is expensive, but a very nice material to use, providing a strong and attractive finish. Silk can be obtained at large model shops, but may be difficult to find in all but the most fully stocked. Apply in just the same way as nylon.

Heatshrink plastic film

The 'modern' material, heatshrink film is a high gloss self-adhesive coloured material, which covers and finishes in one easy step. Certain plastics exhibit a phenomenon known as elastic memory, which means that they can be heated and stretched, then cooled, and once re-heated, they will shrink back to their original size. A shrinkage of 10% is normal. Once stretched, the manufacturer coats the reverse side of the film with a coloured heat-sensitive glue, then protects this surface with a throw away backing of polythene sheet.

In use, the film is cut to size to suit the panel to be covered and, after removal of the backing sheet, tacked into place by melting spots of the adhesive with either a domestic electric iron or special tacking iron. The edges can then be fully sealed with the iron, then the panel heated, either with the iron, or a heat gun similar to a hair dryer, which shrinks the film taut. The resulting high-gloss covering is fuel-proof and reasonably puncture resistant. This method is probably the quickest way to obtain a bright high-gloss finish. It has disadvantages; it imparts little or no extra strength to the airframe, which must be designed with heatshrink film in mind, and it is difficult to seal edges fully. If edged are not sealed, it is easy for fuel residues to seep under the film, damaging the basic model structure underneath the film. Some films are easier than others to use, but it is largely a matter of personal preference however, and none of them are easy to persuade into compound internal curves.

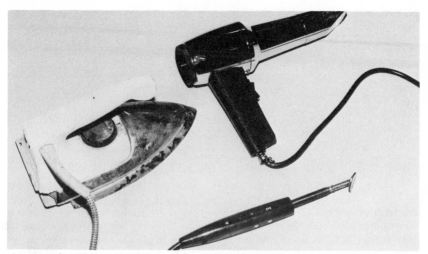

An array of equipment useful for applying and shrinking plastic film. Iron can be used for fixing and shrinking. Heat gun just for shrinking after tacking in place with tacking iron.

Heatshrink fabric

Following on from the development of heatshrink films, heatshrink fabrics seem to be the answer to a prayer, offering the advantages of nylon covering with the benefits of heatshrink plastic film. There has to be a snag, in this case cost. This material is the most expensive of all the options available. Nonetheless many modellers recognise the advantages and are prepared to pay the price.

Most heatshrink fabrics are available in a range of colours, once again self-adhesive with heat-sensitive glue and applied in the same manner as plastic film. This material does bond to internal curves much more readily than other heat-shrink materials.

Glasscloth/resin

Glasscloth/resin finishes are usually only applied to all-sheet models and their use results in a hard, super-smooth surface, which adds considerable strength to the completed model. A specially formulated low-viscosity epoxy resin is used in conjunction with a very lightweight (0.6oz./sq. ft.) close weave glass fibre cloth. The epoxy resin is thinned with methanol (methyl alcohol), then used to paste the glass cloth to the surface to be covered. Once cured the resin can be rubbed down with wet-or-dry abrasive to a super-smooth surface, which can then be painted as required.

Suitability of different coverings

Having outlined all the options, a word about the suitability of each for particular applications is in order. Bearing in mind the type of modelling envisaged, there are several factors which might affect the particular method chosen. Looking firstly at slope-soaring – is the site most likely to be used grass-covered, or is it rocky with clumps of heather? If so, tissue covering and even plastic film will suffer badly on every landing, even good ones! Nylon, or, if the expense can be afforded, heatshrink fabric, will be best.

Thermal-soaring gliders are generally high-aspect ratio (the aspect ratio is the ratio of the wing chord or width to the span) and large span models frequently have the rigidity added by a doped tissue finish, which is essential. Most flying will be carried out from flat grass-covered sites and the low landing speed of the model will help to preserve the covering intact. Coloured tissue (dyed if not available ready-coloured) can be used, which helps keep the weight down, as coloured dopes are heavy.

Power flyers probably have the hardest choice of all to make. Most current R/C powered models are designed with the possibility of the builder using heatshrink plastic film in mind. The structures are going to be adequately strong, but it is almost inevitable that the model will suffer a number of heavy landings while its pilot learns to fly it. Resulting surface damage to plastic film covering and its inherent lack of strength can rapidly result in a model which is severely weakened through fuel seepage and the lack of extra strength provided by covering.

For all its extra difficulty of application, nylon covering is to be strongly recommended, followed by really thorough painting and fuel-proofing. Heatshrink fabric comes a close second, provided it is well painted and fuel-proofed after application.

Whichever material or method is chosen, there is an appropriate procedure to follow on from the basic preparation already outlined.

Covering method for tissue and nylon

Using a reasonable-sized brush – say ½in. – give the whole airframe a coat of full-strength clear dope. Allow this to dry out thoroughly then rub down with glass paper or wet-or-dry paper. Repeat this step in full.

Start covering with the underside of the wing. This is a large straightforward item, usually with minimum complications. Cut a panel of tissue to size with about ½in. overlap all the way round. Now dope the tissue down on to the wing root, smoothing out the wrinkles. Stretch the tissue towards the wing tip and dope it down, stretching it taut spanwise as you do so. Now dope down the leading and trailing edges stretching the tissue taut once again.

Once the tissue is doped into place all the way round, it can be doped over the edges. Where compound curves need to be covered, it will be necessary to make a number of slits in from the edge of the tissue which can then be easily doped over the curve.

Turn the wing over and repeat the operation for the top surface. By starting with the lower surface, if any marks are to be left by the overlap, they will at least be less evident on the lower surface. It is possible to cut through the two layers of tissue together at the overlap and then peel off the waste, leaving the tissue to butt together. When the whole wing is covered, prepare a number of scraps of balsa,

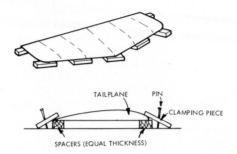

TAILPLANE   PIN

CLAMPING PIECE

SPACERS (EQUAL THICKNESS)

Fig. 22

say 1in. lengths of ½ × ½in. strip, to block the wing up from the building board, as it is advisable to pin the doped wing down flat on to the board to prevent warps as the dope shrinks the tissue. If the strip of balsa is rubbed with a candle before cutting up it will prevent the pieces from sticking to the doped wing. Dope the whole of the wing, then pin down to the board until it is completely dry. Repeat this doping and pinning down process until the tissue has shrunk evenly and has taken on an even sheen with all pores filled with dope. Any unevenness at the edges can be removed by careful rubbing down with 600 wet-or-dry. Use with caution; too much pressure and the tissue can easily be punctured, particularly where an unsupported area meets the edge of sheeting.

Nylon covering should be approached in the same manner, except that the nylon is applied moist and held in place with dressmakers' pins once stretched, until the dope gluing it into place has dried. When the dope is applied to wet nylon it tends to take on an unsightly white appearance. This is known as 'blushing' and usually goes when a second coat of dope is applied. If it does not go, then this probably indicates that the atmosphere where doping is taking place is too humid. A warmer, drier room is needed. It will be almost certainly necessary to rub down the edges of the nylon with wet-or-dry, and a coat of dope over the roughened edges will produce a smooth finish. Carry on applying the dope until the weave of the fabric is fully filled. Do not be tempted to put too much dope on too quickly, as it will run straight through the fabric and form drips on the inside of the covering.

When covering fuselages with either tissue or nylon, cover in panels, overlapping the joins wherever sharp changes of direction occur. For a simple box fuselage, cover the sides first, followed by top and bottom, and trim the material carefully to obtain a neat appearance at the corners.

Covering method for heatshrink materials

Although it is possible to purchase specialist tools for 'heat shrink' covering, an ordinary domestic electric flat iron can be used to good effect. In addition a sharp pair of scissors, a scalpel with fresh blade, a soft duster, a chinagraph pencil and a quantity of cellulose thinners will be needed.

If any filler material has been used it is a good idea to prime the surface before film covering. Proprietary surface preparations are available, or a coat of sanding sealer can be used. If the film is to go over the GRP wing joint reinforcement certainly give this at least one coat of sanding sealer. Identify the backing sheet side of the film and lay the film down backing sheet uppermost. Place the wing down on this and draw round it with the chinagraph pencil, allowing for a ½in. overlap. Cut out the pieces required all in one operation. Set the iron in accordance with the film manufacturers' recommendations and try it for temperature on a scrap of material.

When all is ready, peel off the backing film and commence to tack the film at the wing root. Stretch it towards the wingtip and fix in place. Now stretching the film chordwise, tack it in place at the leading and trailing edges. Apply tension to the film and heat it with the tip of the iron and it will be found that it can be stretched around quite considerable curves at wingtips etc. Trim off excess and seal down the edges and then cover the top surface of the wing. Once the whole wing is covered, the film can be shrunk. Try to float the iron just over the surface without touching it. Once the film has been heated it can be smoothed down with the soft cloth on to the sheet areas as it cools and shrinks. Any traces of adhesive that oozed out from under the edges of the film can be removed with a little cellulose thinners on a soft cloth.

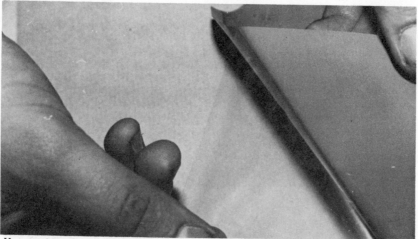

Heat sensitive glue on the covering film is protected by a clear polythene film.

Heatshrink fabric is fixed and shrunk in just the same way, but once the covering is completed a coat of dope followed by fuel-proofer is recommended for power models, and at least two coats of each in the engine and fuel tank compartment. Edges of plastic film can be sealed with epoxy resin glue in the region of the engine bay, which will once again need to be fully fuel-proofed.

Painting and trimming

Dope and tissue or nylon covered models can be simply left in the colour of the material and if a power model, once fuel-proofed, are ready to be flown. Most

Nylon fabric produces a very tough final structure. Material is applied damp, pinned in place, then bonded on and shrunk with clear shrinking dope.

modellers will want to finish off their creation a little more colourfully, however. It may be necessary to give the fuselage and sheet areas of the model a coat of sanding sealer to disguise wood grain or fabric weave fully. Once this has dried and been rubbed down, the surface must be cleaned of dust before painting.

Brush painting is quite adequate, aerosols produce good finishes but are expensive. A spray gun or airbrush is the ultimate and although initially expensive, running costs are moderate. Whatever method is chosen, it must be carried out in a dust-free atmosphere, and proper provision for supporting the parts of the model whilst painting and drying must be made before painting is started. Some form of clothes-line arrangement will suit, though it may be necessary to arrange for fixing a handle or hook of some sort to an inconspicuous part of the model.

Once ready do not attempt to cover the whole model with a single thick coat of colour, but rather use several light coats, allowing each to dry properly before proceeding with the next. If contrasting colours are required, these can be masked off using self-adhesive masking tape. If brush painting, it is advisable to seal the tape edges with clear dope before applying colour over the tape. This is not

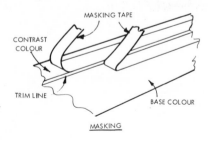

Fig. 23

55

usually necessary when spray painting, a firm rub down with a thumbnail being good enougn. If cellulose paints are used, it will be necessary to fuel-proof the whole of a power model. Many modern epoxy and polyurethane paints are, however, fuel-proof. If in doubt, fuel-proof! Ideally, purchase all the paints, fillers and thinners from a single source and check for compatibility before departing. Most model shop proprietors will know what can be mixed with what, and are a mine of information regarding finishes.

If painted trim is too ambitious, or a quick result is required, there are many self-adhesive colour trim materials available from model shops which are fuel-proof. Use of these materials is particularly to be recommended for applying trim to heatshrink film, as using extra colour film over the base colour can be tricky. Do use a straight-edge and a really sharp scalpel for cutting trim to shape. Make the finish on your model something to be proud of even if using the 'quickest and dirtiest' method around. A well-finished model always lasts longer than one that has had the initial stages skimped.

Simple but effective colour scheme produced with contrasting colours of heat shrink film on this aerobatic slope soarer.

CHAPTER SIX

Installing the Equipment

Now that all the component parts of the model are finished, it can once again be re-assembled, checking to see that such items as elevators and rudders still fit, as paint or heatshrink film may have taken up the original clearance. Once satisfied that all the parts are as they should be, hinges can be fitted to the control surfaces. Several methods are available; all have their places and none can really be considered a 'best' or only way. Two-part pinned hinges of the knuckle type are the freeist, but they are also potentially the easiest type to stiffen up due to either

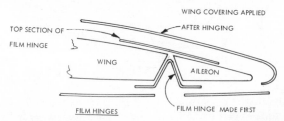

TOP SECTION OF FILM HINGE

WING COVERING APPLIED AFTER HINGING

WING

AILERON

FILM HINGES

FILM HINGE MADE FIRST

Fig. 24

misalignment or penetration of glue or paint. Next on the list comes the plastic film hinge. Usually nylon or polypropylene, these simple hinges are no more than flat pieces of thin flexible plastic sheet which are inserted into slits in both parts to be hinged. Both glue (cyanoacrylate or epoxy) and pinning of some description should be used to keep this type of hinge in place.

Heatshrink film can be used to form hinges, and this method involves covering and hinging in one operation. A definite aerodynamic advantage results from this type of hinge, for when finished the control surface is continuous with the fixed surface. There is no leakage of air through the hinge line and the control surface becomes much more effective. In the same family of hinges moulded flexible plastic types need to be included; these rely on the incredible resistance to fatigue exhibited by plastics such as polypropylene, which can be bent back and forth countless times without breaking along the bend line.

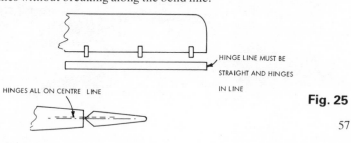

HINGE LINE MUST BE STRAIGHT AND HINGES IN LINE

HINGES ALL ON CENTRE LINE

Fig. 25

Whichever type of hinge is chosen, proper fitting is vital. Too stiff and loads are fed back to the servo, which in turn takes more current, potentially lowering battery voltage and leaving the R/C receiver 'open' to outside interference. Like any chain, the strength is that of its weakest link, and all parts of the R/C installation, hinges included, merit careful attention to detail. Hinge lines must be straight, and all hinges must be in line. Do try not to get waste glue all over knuckle-type hinges; application of a tiny drop of '3-in-1' oil before assembling the parts to be hinged may help. Once the hinges are slipped into place, check by moving the surface to its fullest extent that it will not become hinge-bound. If all is as it should be the hinges can be pinned in place, either with dressmaker's pins or short pieces of cocktail stick. These can be covered over with scraps of covering film, or disguised with a tiny dab of matching paint once fitted. If pushed in from the underside of the surface they should not be too conspicuous.

Control horns can now be screwed into place. Unless the plastic moulding is specifically designed for the purpose do not rely on the fixing screws "self

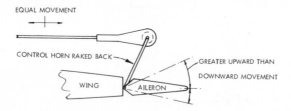

Fig. 26

tapping" into the plastic horns, but instead use nuts on the bolts. Arrange the position of the horn so that the holes for the clevises are exactly in line with the hinge line. This is important, if the holes are to the front or rear of the hinge line a differential movement will be imparted, i.e. for a given movement of the pushrod, the surface will move in unequal amounts either side of neutral. This feature can sometimes be introduced intentionally if more movement one way or the other is needed, as is sometimes the case with ailerons where more 'up' movement than 'down' is desirable.

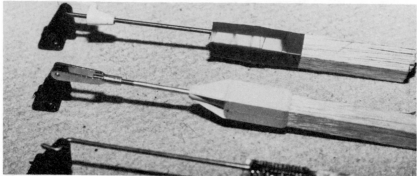

Three different forms of pushrod to control horn connection, coupled with three forms of pushrod construction method.

58

Fitting the servos

The actual installation of the R/C equipment can now be planned, as it will now be obvious where pushrods or other connectors have to run to and how much space is required for the receiver, battery and switch. Once an approximate layout has been worked out, first fit the servos. In all R/C kits provision will have been made for fitting the servos, either hardwood beams or plywood plates being most common. Servos themselves are fitted with mounting lugs in which rubber grommets need to be fitted. These grommets are very necessary, as the internals of the servo will perform best and last longest if protected from vibration, hence the grommets. Select servos which rotate the correct way for the job in hand.

Firstly, plan which side of the servo's output arm or disc the connection to the control surface needs to be fitted to for the straightest, most direct run. Bear in mind that as well as moving backwards and forwards pushrods are likely to move sideways and need a minimum clearance if they are not to interfere with one another. Once the best control run has been established, be reluctant to accept any other. Lay servos in place and switch on the R/C equipment, then operate the controls whilst watching the servos to check that the controls will move the correct way in response to the transmitter commands. It may be necessary to shuffle the servos round until a combination of clockwise and counterclockwise servos is found that suits the chosen installation plan.

Screw the servos down, putting a washer under the head of the screw and tightening the screw up just enough to pinch the washer. The servo will still be able to move about as the grommets flex, but the normal load of the control surfaces should not cause the servos to move much.

The servos have now to be linked to the control surfaces. This can be done in three different ways:

1. Rigid pushrods of hard balsa of not less than ¼ in.sq. section for small models, ⅜ in.sq. for larger models. Pushrods should employ the largest possible percentage of wood to wire end terminations, as the ¼ in.sq. wood is far more rigid than wire of the sort of gauge practical for R/C pushrod connections. The wire ends are usually threaded to take clevises for connection to servo arms or control horns. A 'Z'-type end can be used at the servo, or a simple 90° bend with a moulded plastic keeper. There is something to be said for these latter methods, as it is just possible for a pushrod fitted with clevises at either end to rotate and eventually come unscrewed at one end or the other. Wire ends can be fitted to the wooden section either by binding with thread then coating with glue, sliding on heat-shrink tube, or binding with plastic insulating tape.

2. Flexible cable in tube; the flexible cable can be either a stranded steel type (Bowden cable) or a plastic material. The outer tube is also flexible and is anchored in place at either end and in suitable intermediate positions. Providing that the outer carrying tube is not forced into taking too sharp a curvature, and not in any way kinked, and that the ends are neatly trimmed, the cable will slide reasonably freely. If the cable is not really well installed, these systems can be very stiff. This will result in control surfaces which do not move smoothly and servos which are excessively loaded up. Fixing servo connections and clevises can present problems, as the inner cable is very easy to deform. Either a length of fine gauge piano wire needs to be forced down the centre of

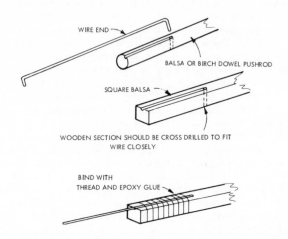

WIRE END

BALSA OR BIRCH DOWEL PUSHROD

SQUARE BALSA

WOODEN SECTION SHOULD BE CROSS DRILLED TO FIT
WIRE CLOSELY

BIND WITH
THREAD AND EPOXY GLUE

Fig. 27

the 'inner' so that the free end is stiffened, or, if it is Bowden cable, tinning with soft solder will do the job. Unless this is done, servo thrust will simply deform the free end of the inner cable without it moving down the tube and moving the control surface.

3. Third of the methods is the push-pull cable, or closed loop system. By far the most positive, slop-free and lightest method, but difficult to adjust for differing amounts of control surface movement. This system is ideally suited for rudder controls on models such as gliders. A double-ended servo arm and double control horn is used with a stranded wire, such as fishing trace, used to link up the horn and servo. The cable can be fixed with short lengths of tubing crimped over with pliers. Adjustment can be built in, but will be needed in each half of the 'closed loop'.

If the model is a kit, once again the hook-up method will already have been determined and instructions will need to be followed regarding use of materials supplied etc. Golden rules for pushrods are to keep them as straight as possible, and as stiff as possible. Make certain that any holes they have to pass through in bulkheads, or where they emerge from fuselage sides, are adequately sized, and that the pushrods do not foul at any point in their travel. Also check that they do not catch up against one another in the confined area at the rear of the fuselage. If they do, re-route them rather than accept a situation that is anything less than 100% satisfactory.

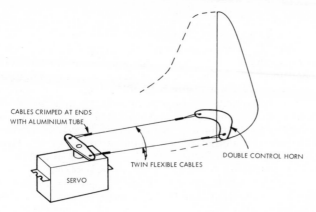

CABLES CRIMPED AT ENDS
WITH ALUMINIUM TUBE

DOUBLE CONTROL HORN

TWIN FLEXIBLE CABLES

SERVO

Fig. 28

Clevises should be screwed well on to the threaded wire ends of the pushrods allowing possibility for future adjustment, but also taking into account the requirement for adequate security at the outer limits of possible adjustment. Once the pushrod installation is felt to be right, a final check is needed to make quite certain that the connection to the servo output arm is not loading the servo excessively at the end of its travel. Move the control stick very slowly, observing the servo output arm carefully. It should continue moving for the full travel of the control stick, plus the extra few degrees provided by action of the trim. If the servo stops before the control stick is moved to full travel, the servo is stalled and will be drawing very high current from the battery – up to 2.5 amps! Correct any such fault that becomes apparent.

Throttle connection

Setting up the throttle connection needs to be approached methodically for guaranteed success. Firstly, connect a single clevis to the throttle end of the linkage, then push the throttle to the fully-open forward position with the trim full forward. Now operate the throttle servo, moving the servo to the fully-open

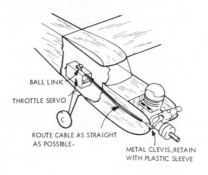

BALL LINK

THROTTLE SERVO

ROUTE CABLE AS STRAIGHT
AS POSSIBLE

METAL CLEVIS, RETAIN
WITH PLASTIC SLEEVE

Fig. 29

position. Select any servo arm hole for convenience and pinch the servo end of the cable between finger and thumb adjacent to this hole. Now move the throttle stick to the closed position, and keeping a firm grip on the cable, pull it to the throttle closed position and see if the point being held still lines up with the chosen hole.

If it does, all is easy – simply solder on the clevis in the appropriate place and complete the hook-up. The chances are that it won't be right, however – there is almost certain to be either too much or too little travel. If possible arrange for the correct travel by selecting a different hole on the servo output arm, further out from the centre for greater travel and vice versa. It is more likely that there will be too much servo travel for the throttle, and it may be necessary to reshape the throttle lever, or drill a further hole in the arm further out. Do not accept a situation which allows the throttle to be either fully-open or closed before the servo travels its full extent.

As the cable length is set up with the trim in the fully-open position, it should be possible to use the full range of trim control to adjust the idle speed and, at fully-down position, stop the engine.

Protecting the receiver

The receiver needs to be well protected against vibration and possible shocks (crashes!). Ideally a rubber, not plastic, foam material should be used. Specially selected material is usually available from hobby shops. Check that the leads from the servos to the receiver are not strained in any way, for if they are, vibration during flight could cause a servo to become unplugged with disastrous results. Some modellers go so far as to retain plugs and sockets with small pieces of insulating tape.

If the space for the receiver is very large, it is worth putting extra balsa bulkheads into the model to keep it in place. This also applies to the battery pack, which in normal circumstances should be packed in a similar way to the receiver and placed in front of the receiver (towards the nose of the aircraft). As the battery pack is quite heavy, a crash can throw it forward and anything in front of it is likely to suffer. Protect the comparatively delicate receiver from this fate.

Fitting switches

Switch fitting needs a little thought, as obviously it needs to be accessible but not in a position where is it likely to become covered in exhaust oil from the engine. Be careful not to place the switch in a position where it may be accidentally knocked to the 'off' position, when the model is hand launched. The underside of the fuselage seems at first sight to be a good position, but it isn't really, since if the model touches down on the switch, it may become damaged or filled with dirt and may even be knocked into the 'off' position during a bouncy landing, resulting in the model actually bouncing back into the air and flying with no R/C guidance possible.

The receiver aerial

Finally the receiver aerial has to be routed out of the model. On no account must it be left coiled up inside the fuselage, nor must it be trimmed to a more convenient length. Either of these actions will result in a serious de-tuning of the receiver and a drastic reduction in range. All may appear to be well at close ranges, but as the model flies away it will rapidly go out of range and out of control. Try to arrange for the routing of the aerial to be well away from the servo and battery/switch wiring, leading it out of the fuselage at the nearest convenient point from the receiver. If necessary, drill a hole and, ideally, fit a grommet to

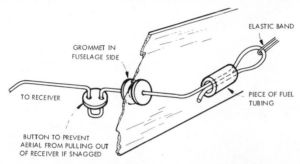

Fig. 30

protect the wire from chafing at the exit point. Additionally, the aerial should be anchored inside the fuselage to prevent an accidental snagging of the aerial wire from damaging the connection to the receiver printed circuit board. Run the aerial to a convenient point such as the tip of the fin and strain it, just sufficiently to prevent it looking too unsightly, with a small elastic band.

If there is an excess of aerial lead, allow it to trail free and do not double it up at the end any more than is strictly necessary for attachment.

Checking the controls

Once satisfied that all the controls move freely to their fullest extent of travel, a full check of the controls should be made. Set all trims to the centre position and, using a straight-edge if necessary, adjust finally all the clevises to achieve exactly neutral setting on the control surfaces. Re-check that all control surfaces move freely to extreme positions.

Control surface movements do cause many questions to be asked. A simple rule of thumb method for setting control surface throws, in the absence of any instructions is the ¼in. rule. Set all throws so that the outer edge of the surface moves just ¼in. On narrow surfaces, the resulting angular deflection will be greater than on wide surfaces, but obviously the effect will be broadly similar. The angles will not look very large but will be quite sufficient for adequate control; they can always be increased if they are found to be insufficient, but too large a movement can make the model impossible to control.

Before putting the model to one side for a full charge of the batteries (and this is essential before any attempt is made to fly) stand behind the model and operate

63

all the controls for a last check that all surfaces move in the correct direction. Pull the elevator stick back – towards the bottom of the transmitter case – and the elevator trailing edge should rise, and vice versa. Move the rudder stick to the right and the trailing edge of the rudder should move to the right. If ailerons are fitted, right hand movement of the aileron control stick should cause the right hand aileron to go up, and left down. Throttle stick forward should open the throttle.

If all is well, plug in the charger and fingers crossed for good weather in the morning!

Three different control connection systems are used here, solid pushrod for elevators, "closed loop" cable for rudder and flexible Bowden cable for throttle.

CHAPTER SEVEN

Flying Powered Aircraft

Before any thought is given to actually going and flying, some thought must be given to where the model can be flown and how the skills involved can be learnt. A chat with the proprietor of the hobby shop where all the equipment was purchased is probably the best means of sounding out local prospects. If there is an active model flying club in the locality, a visit to their flying site, or immediate joining up, is strongly recommended. Membership of a club will almost guarantee access to one or more suitable sites for flying and provide introductions to suitably qualified R/C flying instructors, not to mention the possibility of obtaining the necessary third party insurance through a group purchase scheme. This latter requirement is absolutely essential and must be obtained before any attempt to fly the model is made. It is not expensive, which is a reflection on the generally excellent safety record of R/C models, but accidents can happen! Third party insurance is a universal requirement for members of organised clubs and can be obtained from several sources. Most common for individuals is the MAP scheme – see Appendix.

Club membership is not to everyone's taste, and there will always be those 'lone wolf' types who are determined to shun assistance right from the very outset of their modelling activities; their only contact with other modellers will probably be through the hobby shop where they purchase materials. Lone wolves are strongly recommended to use this contact to check out the local situation with regard to flying sites. The loner will probably have a good idea of where his first attempts at R/C flying are to take place. In his own and other R/C fliers' interests it is as well to have found out just where local enthusiasts fly. Once the information is obtained keep a margin of at least 1 mile separation between the flying sites, otherwise there is a strong possibility of potentially disastrous mutual R/C interference.

It is one of nature's irrefutable laws, in fact the Seventh Law of Cussedness, that whatever day is chosen for first flights of an R/C model, the weather conditions will be unsuitable! That keen sense of anticipation built up during the final day of checks and preparations is all too often too much to be denied and the beginner sets out on a day that is far less than suitable. If help is at hand this may well avert disaster, but for the lone wolf a suitable day is most important. Of the two courses available, self-help or qualified tuition, the latter certainly offers the best chance of immediate success, which does not mean that self-help does not work, given patience and a common-sense approach.

Assuming that the weather is kind, a light breeze, warm, clear sky, dry underfoot, then what next? The model will be fully prepared, batteries which have been on charge overnight will be fully topped-up. Let us first consider what else

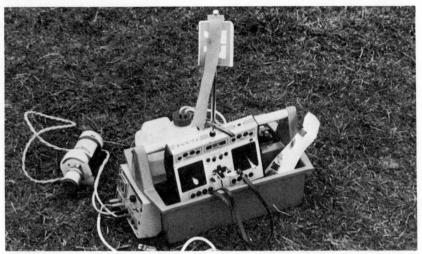

Some form of field box is essential for carrying all the necessary odds and ends out to the flying patch – a moulded plastic "carry all" is one solution.

remains to be done before setting out for that first power R/C flying session. As well as the model and transmitter, rubber bands are needed, pliers, a spanner for the propeller, screwdrivers, fuel, starting battery, glowplug lead – and remember some rag to wipe oily hands on after starting the engine, and to wipe down the model after flights. A comprehensive checklist is a very good idea; spend some of the previous evening writing out the list. Keep it within reasonable limits, for all the equipment including model and transmitter will probably have to be carried some distance out to the flying site.

If introduction to modelling was via an existing club, and help in learning to fly has been offered, or is expected, an arrangement may have been made for a first learning session. If so, all well and good, but if no arrangement has been made and help is definitely felt to be needed do check that the chosen time will coincide with one of the club's regular flying sessions. Many clubs operate quite organised systems of flying training, frequently attaching new inexperienced members to experienced expert volunteers. If the club local to you runs such a system, take advantage of it; many experienced fliers really do enjoy teaching newcomers to fly. Beware of taking someone up on an offer of tuition unless you are confident of his ability, an unlikely situation in well-organised clubs, but on casual generally open sites, such as common land areas, it is possible that the self-styled expert who offers help is little better qualified than you might be yourself. If you have any doubt, stand back and watch the fliers. Then approach one who seems to you to be fully competent. Signs of his competence will be things such as engine always reliable, not cutting out on tick-over or running roughly, landings always good and on the landing/take-off area, sensible flying, no low-level beat-ups over bystanders' heads etc.

Contact prior to actually flying with the club, and discussions with the hobby shop proprietor, will probably have resulted in the purchase of a widely-used brand of R/C equipment which will be of the same mode as that commonly used by the club. If 'buddy box' facilities are fitted to your equipment and it is compatible with that of your instructor in all respects, purchase of a connecting lead is a good investment. The 'Buddy Box' concept allows the two transmitters of the teacher and pupil to be connected together, allowing the teacher 'master' control of the two. In practice, the instructor can perform the take-off and when the model reaches a safe height, can press a button on his transmitter which transfers control to the pupil's transmitter. In the event of any sort of problem, the instructor releases the button and takes over control of the model until a safe situation is regained. The alternative is for the instructor to pass the complete transmitter to the pupil when he feels the time is right, then either reach over the pupil's arms and correct mistakes, or in dire emergency, wrench the transmitter away from the pupil. Different instructors favour different approaches and some distrust 'buddy box' systems, or it may be that your chosen instructor's equipment, or even your own, does not have 'buddy box' facilities. Whichever method of instruction is chosen, you must be prepared to accept the instructor's judgement as to what constitutes a safe situation and one which demands intervention. Models have been crashed as pupil and teacher fight over who is to have the transmitter!

Don't expect too much of the first flight of the model. Remember it will be a first flight of a new engine, aeroplane and R/C equipment and your instructor will almost certainly wish to check every aspect of the complete package fully before handing it over to you. It is essential that the instructor is fully familiar with the characteristics of the model, otherwise he will not be able to judge what does and does not constitute a safe situation while he is teaching. He will also want to satisfy himself that the model is 'trimmed' in the best possible way before passing it over. A poorly trimmed model is no pleasure to fly for the expert and can be a total disaster for the novice. In all probability either the control throws or centres will need minor adjustment; maybe the engine will need some adjustment, so be content to allow the instructor to perform these adjustments. Once the model is correctly set up, the learning proper can be commenced.

Take-off and landing are the most critical parts of the whole flight and most pupils will need some familiarity with the handling characteristics of the model before attempting these phases of the flight. However, there are still those lone wolves around with no instructor to help, so what can they do to help guarantee the best possible chance of success? The following procedure should give both lone wolves and learners under instruction a good idea of what to aim for or what they are being helped to do respectively.

Once at the flying site the model and equipment must be checked over. A definite pre-flight checklist should be performed before each flying session and before each separate flight. The R/C equipment manual will give some instructions about checking, and these should be followed implicitly. Most systems recommend a range check with the transmitter aerial retracted and the model on the ground. Perform this check both with the engine stopped and the engine running. Arrange a system of hand signals with your helper for the engine-on check, which can also be used for a long-distance range check if the equipment instructions require it. A check on the transmitter meter reading must also be carried out to see that it reads within the specified limits. Check number two is for security of all parts of the model, wing bands, tailplane, fin, undercarriage and control surfaces. Also make sure that the propeller is bolted up tight and that it is not damaged in any way; if it is, replace it. Last of all look around to make sure there are no animals or people or other obstructions to which your model could be a hazard or vice versa. When all checks are complete, re-fuel the model, start up, wipe your hands clean of oil, and make ready for take-off.

The ideal method of getting the model into the air is to take off from the ground. A hand launch is a definite second-best, as the pilot has little or no control over the actual attitude in which the model leaves the launcher's hand. If the hand launch is a good one, the model will fly smoothly away, but if released at a nose high or low angle, or with one wing dropped, the instant reaction necessary will almost certainly not be there. If a hand launch is the only possibility, find someone to perform it, so that you are able to keep both hands on the transmitter, ready for instant correction of incorrect attitude.

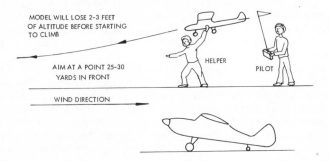

MODEL WILL LOSE 2-3 FEET OF ALTITUDE BEFORE STARTING TO CLIMB

AIM AT A POINT 25-30 YARDS IN FRONT

HELPER

PILOT

WIND DIRECTION

Fig. 31

The model should be launched at flying speed. What is flying speed? Generally a brisk trot into wind terminating in a firm forward push should result in the model flying away. Avoid the temptation to run at full tilt, as the result is likely to be an uncontrolled heave of the model. Wings should be level and the model aimed at a point on the ground 20-30 metres in front of the launch point. Arrange a good clear signal for your helper to start on his launch run and once final checks of control surface movements have been made, give the signal. Following a good launch, the model should be allowed to build up speed in a very shallow climb. Do not attempt to pull the elevator control stick back in an attempt to gain the safety of altitude. Speed can be converted into altitude, altitude can be converted into speed. No speed, no altitude spells danger. After a few seconds of flight the control stick can be gently eased a little way back to give 'up' elevator and a gentle climb initiated. If a take-off is possible, there are two different techniques, depending on the type of undercarriage fitted to the model. If the model has a tricycle or nosewheel, point the model exactly into the wind, and ask your helper to restrain it whilst you check the controls. Open the throttle fully, give the signal for release and allow the model to accelerate. It may be necessary to correct its heading with the rudder, but use only as much control as is necessary. If the take-off run goes badly wrong, shut the throttle and try again; do not attempt to lift off unless the model is facing into the wind. Once the model seems to be

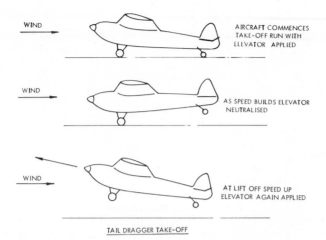

WIND

AIRCRAFT COMMENCES
TAKE-OFF RUN WITH
ELEVATOR APPLIED

WIND

AS SPEED BUILDS ELEVATOR
NEUTRALISED

WIND

AT LIFT OFF SPEED UP
ELEVATOR AGAIN APPLIED

TAIL DRAGGER TAKE-OFF

Fig. 32

moving fast, ease back the elevator control stick and the model should lift off smoothly. Use plenty of runway and don't lift off until nearly at the end, once again remembering the speed/altitude equation.

A tail wheel or 'taildragger' type of model requires a different technique. It is necessary to hold on some 'up' elevator as the throttle is opened, otherwise there will be a tendency for the model to nose over on to the propeller, stopping the engine. Once released, the model will gain speed with the tail still on the ground, and when fair speed is attained, ease back on the elevators which will allow the tailplane to lift. A little more speed and a gentle application of 'up' elevator will cause the model to rise smoothly off the ground.

Do allow the model to gain plenty of altitude (100-150ft) before making any turn, which should be initiated smoothly with the rudder coupled with a small amount of 'up' elevator to prevent the nose dropping during the turn. Only use as much control movement as is necessary to steer the model gently around within 200-300 yards of the launch spot. If the model is gaining too much height, throttle back the engine a little rather than push the elevator control stick forward. This would cause the model to speed up rapidly in a full power dive, and when the control stick is released the model would zoom back skywards, needing further 'down' elevator to prevent a stall; the model would then speed up etc.

Do not attempt to land the model under power for the first time. Wait until the engine runs out of fuel, then glide a circuit or two round the field, depending on altitude, and concentrate on keeping the model directly into wind for the touch-down, which will keep actual speed over the ground to a minimum. Just before the moment of touchdown, feed in 'up' elevator to flare out the landing. The model should just be on the point of stalling as it touches down, in other words it should stop flying just before it touches the ground.

When the excitement is over, take stock of what happened. The model flies and is down in one piece, but the success of the flight was probably more a result of good design of the model than expert piloting. Almost certainly, the model flew

69

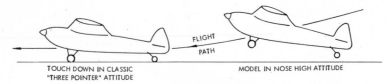

TOUCH DOWN IN CLASSIC
"THREE POINTER" ATTITUDE

MODEL IN NOSE HIGH ATTITUDE

THREE POINT LANDING WITH

TAIL DRAGGER

TRICYCLE UNDERCARRIAGE

TOUCH DOWN

MAIN LEGS SHOULD   CONTACT FIRST OTHERWISE
MODEL IS BOUND TO BE   FLYING TOO FAST

Fig. 33

where it wanted to go, with intervention from its 'pilot' when the situation showed signs of getting out of hand. Subsequent flights, with less adrenalin flowing, should firstly give increasing familiarity with the model, and gradually a tendency to fly the model in the direction you want the model to take, rather than the other way round.

Of course, the foregoing description is probably idealised and reality may be a little more fast-moving and eventful. It is as well to be prepared for a model which may do one or more of several things, and maybe all at once. It is not entirely impossible for the model to fly perfectly straight from the first launch, but the chances are that the elevator or rudder will not be adjusted exactly correctly. This maladjustment can result in the model veering away to one side or the other on the launch, or climbing away too steeply, or diving at the ground. Assuming that the 'lone wolf' is quick enough with remedial action on the control sticks, and the model does manage to fly away from the launch, and providing the problem is not too severe, gain altitude, then correct the situation with the transmitter trims. If the model veers away to the right, move the rudder trim over to the left and vice versa. Likewise correct a climb or dive with elevator trim. If trim adjustment will not cope, shut the throttle and attempt a straight into-wind landing. If this is even 400 yards away, don't worry, a walk is better than a smashed model resulting from the attempt to control a badly-out-of-trim model.

Before any further flights, adjust the control surface clevises to correct the out-of-trim situation, and don't forget to set the transmitter trims back to neutral once the adjustments are made to the model. Now that the model is properly adjusted, subsequent flights should be made with a definite aim in mind. Positioning the model with precision is necessary if good landings are going to be made. Start by flying the boundaries of the field first one way round, then try the reverse direction. Once this has been mastered, try flying a figure-8 course with the intersection of the '8' over the take-off area. This will highlight the problem of rudder reversal, as when the model flies towards you, you will find that the effect of the rudder appears to be reversed. Try to cope with this without turning and watching the model over your shoulder. One good way of coping with the problem is to imagine that you are using the rudder control stick as a prop under the wing that drops. As the model turns, one wing, that on the inside of the turn, drops.

70

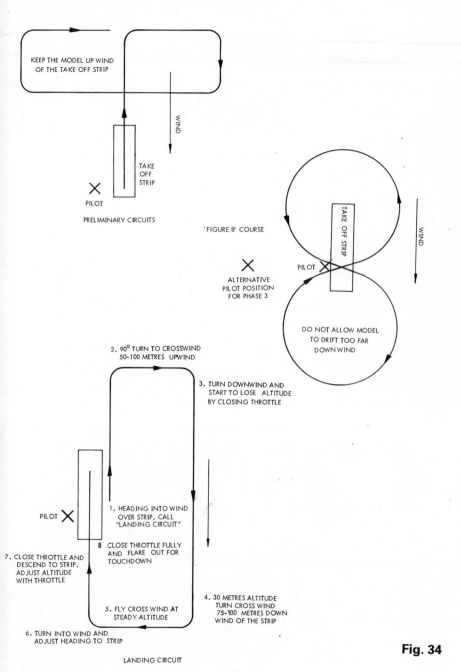

KEEP THE MODEL UP WIND
OF THE TAKE OFF STRIP

WIND

TAKE
OFF
STRIP

PILOT

PRELIMINARY CIRCUITS

'FIGURE 8' COURSE

TAKE OFF STRIP

WIND

PILOT

ALTERNATIVE
PILOT POSITION
FOR PHASE 3

DO NOT ALLOW MODEL
TO DRIFT TOO FAR
DOWNWIND

2. 90° TURN TO CROSSWIND
50-100 METRES UPWIND

3. TURN DOWNWIND AND
START TO LOSE ALTITUDE
BY CLOSING THROTTLE

PILOT

1. HEADING INTO WIND
OVER STRIP, CALL
"LANDING CIRCUIT"

8 CLOSE THROTTLE FULLY
AND FLARE OUT FOR
TOUCHDOWN

7. CLOSE THROTTLE AND
DESCEND TO STRIP.
ADJUST ALTITUDE
WITH THROTTLE

4. 30 METRES ALTITUDE
TURN CROSS WIND
75-100 METRES DOWN
WIND OF THE STRIP

5. FLY CROSS WIND AT
STEADY ALTITUDE

6. TURN INTO WIND AND
ADJUST HEADING TO STRIP

LANDING CIRCUIT

Fig. 34

71

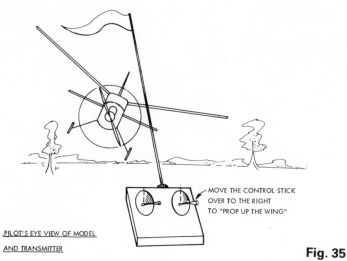

MOVE THE CONTROL STICK
OVER TO THE RIGHT
TO "PROP UP THE WING"

PILOT'S EYE VIEW OF MODEL
AND TRANSMITTER

Fig. 35

When the model is flying towards you, move the stick over as if to prop up the dropping wing.

When flying the model towards you can be coped with, landing approaches can be attempted. Try these at altitude first, commencing the landing approach via a positive rectangular flight pattern around the field. The first leg of the pattern should be directly into the wind over the landing area, gradually descending through the next three sides until on the fifth leg, into the wind once again, the model is correctly lined up on the landing area for touchdown. Control the rate of descent with the throttle and be prepared to open up and try again if too high. Do not push the elevator stick forwards to lose height, since this will just increase speed. Speed can be reduced by pulling the stick back, but beware of trying to fly too slowly or the model will stall. You see why approaches should be attempted at altitude first. Don't be too disappointed if landing on the take-off area proves impossible, since it is far better to concentrate on arriving at ground level at the right speed in the right attitude, and into wind. The spot landing will prove simple eventually. All of the problems caused by an out-of-trim model, disorientation, control reversal and correct judgement of landing approach height are a matter of experience, which is much more easily gained in the presence of an experienced R/C pilot. For the least painful introduction to R/C flying, club membership and competent instruction cannot be recommended strongly enough.

If you are determined to go it alone, the best of luck; if things go badly wrong the very best thing to do is to shut the throttle, let all the controls spring to neutral, and leave the model to sort itself out on its own. A well-designed trainer model will almost certainly recover from any attitude, however extreme, providing always that it is at a sufficient altitude. Once the model has steadied down, open up and try again. The experience gained in learning unaided is irreplaceable, the sense of satisfaction immense, the heartbreaks almost unbearable!

CHAPTER EIGHT

Flying Helicopters

Learning to fly a helicopter is almost always a solo affair. Advice can be given, and help in initially adjusting the model, but 'buddy box' methods, and direct intervention by an instructor, are just not feasible. Before a helicopter can be flown around, the would-be pilot has to learn to hover the model, as every flight begins and ends in the hover. As hovering is carried out close to the ground, there is insufficient time for an instructor using a 'buddy box' to intervene and correct a potentially dangerous situation before the model goes totally out of control. Therefore the helicopter learner is almost forced to try for himself. Providing proper caution is exercised, this is no bad thing. Helicopters demand a high level of co-ordination between the four controls, and this can only be developed by long experience which can be gained with the model just skimming the ground and at very low level. Helicopters are complex mechanical devices and need to be properly maintained if they are to fly at all and be controllable. Do not be prepared to accept balance that is 'nearly right', or trim that is a 'little out' or an engine that is 'fairly reliable'. Follow instructions to the letter with regard to preliminary adjustments and make certain that the engine performs faultlessly.

Start off by fully familiarising yourself with what the controls do. Throttle controls the helicopter's vertical speed up-and-down in the hover, and forward, backwards and sideways speed once the model is flying around. The tail rotor pitch control compensates for variation in engine torque as the throttle is used. The tail rotor pitch needs to be adjusted continuously as the throttle is operated to

Author's Kalt helicopter, equipped with a 0.45 cu. in. motor. This model has collective pitch and autorotation facility.

73

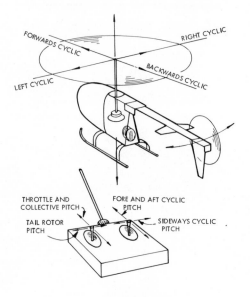

Fig. 36

keep the model facing in the correct direction. Cyclic control is used to move the helicopter forwards, backwards and from side-to-side. All of these controls are interactive, not mechanically but dynamically whilst the helicopter is flying.

When ready, fuel up, start up, and stand back from the model, 5 yards to the rear and left-hand side. Open the throttle and as the rotors speed up, try moving the tail rotor pitch control and watch to see what happens. Watch the nose of the model – right-hand pressure on the tail control moves the nose to the right and vice versa. As the rotor speed increases and the model starts to 'swim' on the ground the tail rotor will become progressively more positive. Keep adjusting the throttle so that the model is 'light' but not lifting off; run at least two tanks of fuel through the engine during this stage of familiarisation and if necessary more, until the tail rotor control is automatic and the nose of the model can be pointed in any direction at will, and brought precisely to any required direction without second thought.

As this area of familiarity progresses, some attempt should be made to get the feel of the cyclic controls. Gentle movements of the stick whilst the model is 'swimming' will cause the model to rock on its skids. Take care, as too fierce use of the controls will cause the model to tip over. Take at least two to three hours to complete this phase of learning then gradually increase the rotor r.p.m. until the model actually lifts clear of the ground. Providing the lessons with the tail rotor have been well learnt, there should be no problem in keeping the model facing into the wind. It may seem at this point that the tail rotor trim control is not correctly set, but beware of altering this, for whilst the fuselage of the model is still in the turbulence caused by the rotor downwash striking the ground, it is difficult to set the trim correctly. Do set some forward cyclic trim, just enough to cause the model to creep gently forward as it clears the ground.

It is much easier to control the model if there is a light, steady breeze. The fin on the end of the tail boom will have a weathercock effect, keeping the model heading in a constant direction into the wind. If things go badly wrong, beware of shutting the throttle violently on a collective pitch model, as this action can cause the rotor blades to drop downwards smartly and hit the rear fuselage or tail boom. Instead, attempt to keep the model upright by use of the cyclic controls and, if absolutely necessary, ignore the tail rotor whilst attempting to set the model down. It is safer for the model to touchdown, even whilst 'pirouetting' than to touch down heading in the correct direction, rotor tip first!

When the model can be controlled just clear of the ground, a little extra throttle will see the model rising quite rapidly. Delicate use of the throttle is needed to maintain a constant height and as forward trim has been set, the model will tend to move off into the wind. Follow the model and perform a number of 'hops', terminating each one by slight rearward pressure on the cyclic control stick combined with a reduction in rotor r.p.m. The model should stop forward movement and sink towards the ground. Just before touchdown quite an increase in power is needed as the rotors move into the very turbulent ground wash, as well as a touch of extra pitch on the tail rotor to stop the increased torque swinging the model round.

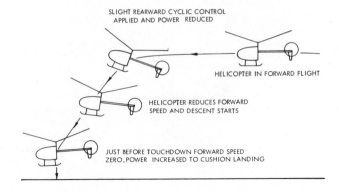

SLIGHT REARWARD CYCLIC CONTROL
APPLIED AND POWER REDUCED

HELICOPTER IN FORWARD FLIGHT

HELICOPTER REDUCES FORWARD
SPEED AND DESCENT STARTS

JUST BEFORE TOUCHDOWN FORWARD SPEED
ZERO. POWER INCREASED TO CUSHION LANDING

Fig. 37

75

After a couple of dozen or so of these hops have been safely accomplished, reduce the amount of forward trim. This will mean that instead of the model lifting off into forward flight, there may be a tendency for it to drift backwards with the breeze. Correct this tendency with forward cyclic control, likewise any sideways drift. At this point the helicopter is genuinely hovering. Not until hovering is mastered fully is it really safe to attempt circuits around the flying site. Flying circuits, figure-of-eights or rectangles is child's play compared with hovering. When the model is raised to a reasonable height (10-20ft) in the hover, forward cyclic control will cause the nose of the model to dip, and the model to accelerate quite rapidly. More often than not, once the helicopter is moving at reasonable speed into the wind, no further use of the tail rotor control need be made. Cyclic control will cause the model to turn, climb and dive. Once the first circuit is complete, make a proper landing approach into the wind, throttling back slightly to cause the model to start to descend. As the landing spot approaches, the throttle setting will need increasing to slow the rate of descent whilst simultaneously correcting any swing with tail rotor control and also pulling back on the

Power plant for helicopter with cooling fan and additional cylinder finning required for sustained high speed running

Author's helicopter – a Kalt 50 Baron, in hover just after lift-off.

cyclic control to arrest the forward motion. If judged correctly, the model should stop moving forward and be hovering about 10 feet above the touchdown spot. It then becomes simply a matter of repeating the by now familiar let-down procedure.

Once the model is hovering, sideways flight, either way, and backwards and forwards flight can be used for exact positioning. After each short flight, check over the essentials, fuel level, correct free motion of all controls, all nuts and bolts tight. There are so many things that can go wrong that regular thorough checks are very necessary. Be careful that the operating time limit of the R/C system is not exceeded, since a lot of flying hours can be logged up whilst learning to fly a helicopter and flat batteries spell total disaster.

Never attempt to fly unless the engine is performing faultlessly; any signs of erratic running or overheating must be investigated and cured before flight is attempted. If looked after and adjusted properly, engine failures are very rare. Good quality fuel, correct mixture strength and a reliable filter are musts. If the engine does fail in the hovering phase, there is little that can be done. However, failure during a circuit does not have to be disastrous, particularly if a free-wheeling rotor is fitted to the model and the rotor blades are set to run to negative pitch at the minimum throttle setting. If the engine does fail, close the throttle immediately, which will put the blades to full negative pitch and allow the air rushing through them as the helicopter drops to drive them round. Push the cyclic control stick forward so that the model builds up forward speed. Just before the model reaches the ground, 3-6ft altitude, push the collective pitch control into positive, when the inertia of the spinning rotors will be converted into just enough lift to cushion the landing. Once mastered, autorotation, as it is called, can become an exciting party trick.

The foregoing is not intended to be a complete manual on flying an R/C helicopter, more as a guide to what is to be expected and an attempt to show that R/C helicopters are not untameable, nor filled with mysteries. They are regularly flown and enjoyed by thousands of modellers, who enjoy the unique ability of the helicopter to hover motionless over a spot and ascend and descend vertically at the pilot's will.

Flying in confined spaces is an advantage of helicopters – once the pilot is sufficiently expert! (Graupner photo)

CHAPTER NINE

Flying Gliders

In previous chapters, some brief mention has already been made of slope-soaring and thermal-soaring gliders. Both are essentially simple models, having no engine and fuel system to worry the novice, and only, at their simplest, a two-channel R/C system. Once the decision to ignore the noisier side of R/C flying has been made, a simultaneous decision regarding slope or thermal soaring needs to be made, for both have their special needs.

Slope-soaring

To slope-soar, a suitable site is needed – a slope, but not any slope will do. There are several basic criteria to be met before a site can be said to be suitable, especially where the novice is concerned. Prevailing wind direction is one such criterion. The UK experiences mainly westerly winds, and if the only available hill th ɩ is within reasonable distance faces east, occasions when it can be used will be f v and far between, in inland areas. Coastal regions are more fortunate, as the e ning inshore breeze phenomenon can make slopes very close to the coast usable more regularly, whichever direction they face. Ideally, for regular slope flying, slopes facing north, south, east and west should be within easy reach.

For the novice, the slope should be very open, both the slope face, and the top area. Trees and even large bushes on the face of the slope generate turbulence and present serious hazards to the model if a slope-face landing becomes necessary. A clear landing area is very important. Expert fliers are able to put their models down even across the slope and in very small areas. The novice really does need a clear unobstructed area to allow less than precise positioning of the touchdown. Lack of obstructions that will generate turbulence and the direction the slope faces are more important than the angle of the slope or the height; 30° angle and 100-150 feet height will give good lift. It is quite possible to fly a slope-soarer from sites such as sand dunes of only 25ft height, but the area of lift is generally quite small and flying skill of a high order is needed. Seaside cliffs are good soaring sites, but experience is needed, for mistakes can be costly, at the least an exhausting climb to the foot of the cliffs and back, at worst a lost or damaged model in the sea. Finally on the list of slope criteria comes access. If hiking figures high in your list of interests, this may mean that a wider variety of sites is available to you. If you feel that more than ten minutes' walking or climbing is too high a price to pay, perhaps slope-soaring is not for you. Very few sites are accessible directly by car, most involve a walk, some a very strenuous climb, carrying models, transmitter,

79

*A wide array of model types at this slope soaring site during a competition.
(Photo B Gowland).*

picnic lunch etc. Most ardent slope-soarers could earn a good living as Sherpas after a year or two's model flying!

What exactly keeps the model in the air? An eminently reasonable question. Quite simply – gravity! If a two-pound weight is dropped from a cliff it falls straight down, accelerating under the force of gravity until it reaches the ground. If a model glider is launched forward through the air, gravity causes it to descend, but due to its aerodynamic properties, the acceleration caused by gravity can be directed so that the model moves forward as well as down. Forward motion of the wings through the air produces lift – the model is flying. It can only fly for as long as it can descend, however. When ground level is reached, there is no way it can reverse the downward gravity-induced tendency and climb back up again. This is the situation in still air, the situation that the thermal-soarer, or slope-soarer with no wind, finds itself in. Given a steady breeze on to the angled face of a slope, then the situation changes. Instead of the model descending through still air, the angle of the slope deflects the air current upwards and thus the model is launched into air that is rising. Providing the angle that the slope deflects the air upwards is greater than the glide angle of the model, the model will rise in the rising air currents, flying above the launch point in many instances. So although gravity is attempting to cause the model to descend, this force converted into forward motion causes the model to fly and even climb in the slope lift.

Coupled with the lift, there is almost always turbulence behind the lip of the slope. If the slope top is a sharp ridge, the turbulence can be very strong, snatching the model out of control and smashing it into the lee of the slope if the pilot is not ready to take instant correcting action. Slopes with flat, or gently sloping, lee-sides are best; a large flat area on top of the slope will give only minimal turbulence making it ideal for the novice.

For first slope flights, do make sure that the wind is blowing exactly on to the face of the slope and is not too strong. Ideally ask advice from an experienced flyer who knows the slope you intend to fly from. Every slope has different characteristics which need to be learnt, so that local knowledge does matter. The exact type of model you have chosen will dictate the wind strength range inside which first flights can be attempted. Too strong a wind and flying will become a battle even

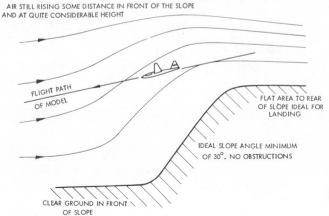

AIR STILL RISING SOME DISTANCE IN FRONT OF THE SLOPE AND AT QUITE CONSIDERABLE HEIGHT

FLIGHT PATH OF MODEL

FLAT AREA TO REAR OF SLOPE IDEAL FOR LANDING

IDEAL SLOPE ANGLE MINIMUM OF 30°. NO OBSTRUCTIONS

CLEAR GROUND IN FRONT OF SLOPE

Fig. 38

81

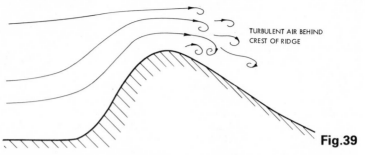

TURBULENT AIR BEHIND
CREST OF RIDGE

Fig.39

for the expert, too little and repeated walks up and down the slope will be the order of the day.

If there is a slope nearby which is regularly used by local flyers, do make contact with them and ask for help. One of their number is almost sure to offer to check over the model and help with the first flights. At the very least, ask the proprietor of the shop where you purchased the whole outfit to give the model a check over. Beware of just tramping on to a nearby convenient hill – all land in the UK belongs to someone, and it may be that even if other modellers have been seen using it, they do so under strict conditions. You could prejudice their use of it by thoughtlessness.

Slope right, wind right, model right, now do you launch the model? Not yet, make sure before switching on for the first time that the frequency is clear. If a peg system is in use, obtain the correct peg from the board, if not shout loud and clear before switching on "Switching on Blue" (or green etc.) and pause before making the final step in case you have overlooked someone. Range-check the system in accordance with the makers' instructions then head for the edge of the slope. Lift and turbulence are slightly less below the lip of the slope – it may be as well to move down the face of the slope a few yards before launching. Check all the controls and launch the model firmly into the wind away from the slope. Providing the model does not rear up violently or perform any other untoward action it should be possible just to watch it head away from the slope, gaining height steadily. If some action is needed, use only as much movement of the control stick as is needed to correct. Too much movement will cause the model to over-react, requiring corresponding further correction. Allow the model to fly at least 50 yards out from the slope, or until it shows signs of no longer climbing. Any signs of loss of height will probably indicate that the model has flown outside the band of lift and will need turning to bring it back into lift.

Turn either to the right or left, it doesn't matter which way, though obviously if you are flying at the extreme left hand end of the slope, a right hand turn is necessary. Smooth positive control input is needed and should be held on until the model has turned nearly 90°. Release the control stick just before the model is parallel to the slope. Let the model drift along parallel to the slope and when circumstances dictate, turn away from the slope into the wind. On no account turn towards the slope. After a 100-yard beat along the slope, the wind will probably have caused the model to drift quite close in to start with. Once turned towards the slope the model will be travelling with the wind, and will move into the slope very quickly and a crash is the almost inevitable result of the inexperienced flyer turning towards the slope.

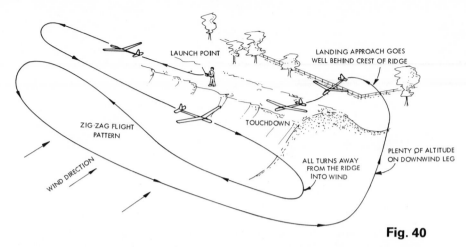

LAUNCH POINT

LANDING APPROACH GOES
WELL BEHIND CREST OF RIDGE

ZIG-ZAG FLIGHT
PATTERN

TOUCHDOWN

PLENTY OF ALTITUDE
ON DOWNWIND LEG

ALL TURNS AWAY
FROM THE RIDGE
INTO WIND

WIND DIRECTION

Fig. 40

Continue the beating up and down the face of the slope, always turning away from the slope at the end of each beat. If too much height is being gained, put the nose of the model down into a shallow dive, or fly out from the slope into the weaker lift area. Flying the slope in this way can be continued until either the R/C batteries run flat or the wind drops, but eventually a landing becomes essential. Before turning downwind to start a landing approach, gain plenty of height, as the downwind turn back into the wind can lose a lot of height, and there will be no slope lift to the rear of the slope. The altitude will have to be converted into forward speed to reach the actual touchdown point chosen. Speed is also a bonus if the air is turbulent behind the slope face, helping the model combat the rough air and remain much more controllable. The most frequent landing error is to make the behind-the-slope turn too low, preventing the model from being brought up to the chosen landing point. If the other extreme is reached, the model can always be flown straight back out to the lift again before a second attempt is made.

Do make absolutely certain that all other flyers and any bystanders know what you are about to attempt – shout "Landing!" loudly before your final approach and "Overshoot" if you are forced to go round again. Always put the safety of bystanders first – if hitting a bystander or crashing are the only options open, crash the model.

Thermal-soaring

Thermal-soaring gliders present a different type of challenge. In spite of many modellers' thoughts to the contrary, thermals – the rising currents of air that the thermal-soarer is designed to utilise – do occur all the year round. All the pilot has to do to prolong the flight beyond the five or so minutes possible from a tow-launch is to find them! But this is probably progressing too far; problem number one is how actually to get the glider flying from a flat-field site.

A hand launch is possible, but flight will only be of a few seconds' duration. Taking a leaf from the full-size glider fliers, the R/C model is generally launched by towing to altitude on a long tow-line, typically 150 metres, and then releasing

83

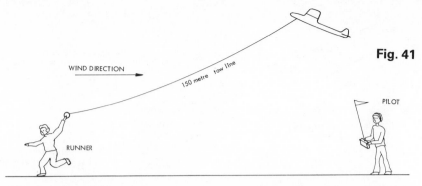

Fig. 41

WIND DIRECTION

150 metre tow line

PILOT

RUNNER

to glide back to the ground. Unlike the power machines used in full-size gliding, we use the simpler solution of keeping the line at full length and using a runner (an athletic friend!) to tow the model. Using this method, the model climbs very steeply, 45° angle of climb or greater, until almost over the head of the person towing. The pilot needs to do very little in terms of controlling the model whilst it is on the towline, since the towing hook is carefully positioned so as to cause the model to adopt the correct climbing attitude and also to track straight. In practice 150 metres of nylon monofilament fishing line of 30-40lb breaking strain is stored on a drum of some sort. This drum is fitted to a geared winder for rapid recovery of the line after launch. On the end of the line a metal ring and a brightly coloured pennant are attached. The ring hooks over a tow-hook on the underside of the glider fuselage whilst the pennant serves two purposes, (1) to help drag the line away from the model when the tower stops his run and (2) to allow the pilot to see that the line has released from the model. Reputable kit models will have the tow-hook positioned correctly, providing the model is properly balanced on the balance point indicated. The line is attached to the hook (it must be free to fall off) and the line run out directly into wind. On a pre-arranged signal, the tower starts to run into wind. As soon as a reasonable tension is built up in the line, the model is released in a climbing attitude with a firm push, up and forward. The model should enter a steep climb straight away, with the rate of climb governed by the running speed of the tower and the strength of the wind. If a stiff breeze is blowing, almost no running will be required as the model will 'kite' to the top of the line. If there is only a light breeze or a flat calm, it may not be possible even with a very athletic tower to obtain very much altitude.

When the model is at its maximum altitude, it will automatically level off and no amount of running will persuade it to get any higher. By this time the tower will probably have realised he is wasting his efforts and stopped running. In calmish

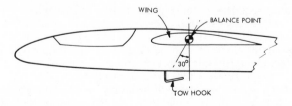

WING

BALANCE POINT

30°

TOW HOOK

Fig. 42

conditions the model will now self-release as drag from the line and pennant pulls it away from the hook. In windy weather it may be necessary to dive the model to slacken off the line sufficiently to allow it to blow off the hook.

A very popular method of launching single-handed is the 'Hi-Start' or Bungee. For this launch method a length of elastic, either cotton covered like the luggage straps used for car roof-racks, or a rubber tube known as surgical rubber tube, is attached to a shorter length of line. The elastic portion is anchored to the ground upwind of the launch point, and the line is stretched in accordance with the Hi-Start makers' instructions. The model is hooked on to the line and released just as for a hand tow launch and flown and released in an identical manner. Hi-Start launches are only really effective when a breeze is blowing, as flat calm days do not allow the model to kite to altitude and all the energy of the bungee is used in just moving the model forward through the still air to maintain flying speed.

A third possibility, although to mention it in thermal-soaring circles might raise a few eyebrows, is to use an ancillary power pod fitted with a small capacity I.C. engine. To many glider enthusiasts engines are absolutely 'out', but as a safe means of getting a thermal-soarer to soaring altitude, they have much to recommend them for the novice, particularly the lone hand. Single-handed tow launches are not very feasible, bungee launches rather awe-inspiring as they are very much akin to catapult launches, whereas the small-sized engines used for power pods provide a very slow forward speed and a gentle rate of climb. Things happen slowly, if they happen at all, giving the novice reasonable time to apply the right corrective control movement. A typical engine size for a 100in. span glider would be 0.10cu.in. (1.6c.c.) coupled with a 1oz. fuel tank. The engine need not be a high-performance type, an ordinary 'sports' type motor being adequate. Power-assisted gliders are rarely expected to take off, or even capable of it, and they will need to be hand launched. Before climbing, a fair distance is needed for the model to accelerate.

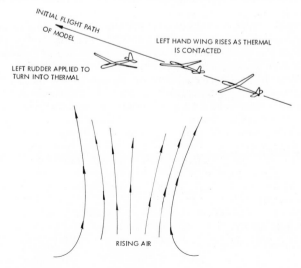

Fig. 43

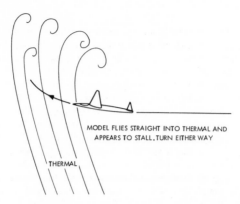

MODEL FLIES STRAIGHT INTO THERMAL AND
APPEARS TO STALL, TURN EITHER WAY

THERMAL

Fig. 44

There are other variations of the above methods of launching; power winches and pulley systems for hand tow are possibilities but beyond the scope of this book. Once the model has been fully prepared for first flights, all the previously described precautions for power and slope-soaring flight need to be followed with regard to flying site, frequency discipline and pre-flight checks. Except for power-assisted models, launching requires reasonably experienced assistance, but once the model is actually at altitude and released from the line, it will be the easiest type of model to control. Control response is gentle, speed is not high, and the model lands at its flying speed with no intervention required apart from steering it down on to the chosen spot.

There is, of course, more to flying a thermal-soarer than gliding steadily down from a 500ft altitude launch, for the object is after all – thermal *soaring*. As already mentioned, thermal upcurrents can occur all the year round, although most easily detected and used during warmer spring and summer months. Air, and the currents in it, are invisible, so how can the presence of the thermal be detected? Quite simply, the model is its own thermal detector. Assuming that it is correctly trimmed, i.e. it flies in a steady straight flight pattern, any thermal that it crosses or meets will cause the pattern to be altered. If the model flies straight into a thermal the nose will rise and it may even stall. If just one wing is caught, then it, too, will rise. Using a thermal discovered by observation of a wing rising, usually indicated by an unexplained or unwanted turn, is probably easiest. Simply turn towards the wing that lifts. If the left wing lifts, a turn left will bring the model into the thermal, whereupon the model will probably be seen to rise noticeably. A thermal encountered head-on is more difficult to cope with, for a turn either way may or may not centre the model in the thermal and only experience can tell whether the action was the right one. Once in the thermal, it is usual practice to circle, drifting downwind with the thermal until sufficient altitude is gained.

Once at altitude, if the thermal is very strong, coming back down again can become a very real problem. The very last thing to do is push the nose down and dive. It is all too easy to misjudge speed and angle at altitude and over-stress the model to the extent that it breaks up in the air. Firstly, try steering the model upwind out of the thermal, or, if this does not seem to be working, try stalling the model by holding on full 'up' elevator; do not allow the stalls to develop into a violent switchback, however. Maybe a flat spin is the answer – pull back the elevator steadily until the model is on the point of stall and then feed in full rudder control to start the spin. Recovery at a safe altitude should be simple, just release the controls and be ready to correct any zoom and stall that might result. Consecutive loops can work, even flying the model upside down, when its efficiency is very poor, causing it to descend.

Best of all are spoilers, which operate by pushing vertical plates out of the wing along the chord line which destroy or spoil all the lift along the length to which they are fitted and also generate a lot of extra drag. Spoilers do, of course, need an extra control function for operation and also quite a fair amount of additional constructional and installation work, but for the sports or open competitive flyer they are recommended.

Every thermal upcurrent seems to be accompanied by a downcurrent, in soaring terminology "sink". This can explain why a model unexpectedly produces a shorter flight than usual. If sink is encountered, fly rapidly away to a totally different area of the flying site, looking out for lift on the way, of course! Apart from using the model as a thermal detector, experience and observation of local conditions can help. Birds frequently fly in thermals, and a flock of seagulls circling at altitude is a sure sign. Strong thermals will pick up scraps of paper and whirl them into the air. It is even possible to feel the temperature change in the atmosphere as a warm thermal bubble passes by. It can be noticed that thermals seem to occur at specific points on any particular site more regularly than others. Flat roofs on factories, large housing estates, stubble fields in among grasslands are good thermal sources. Careful observation of local and weather conditions and experience are the best way to develop a 'nose' for lift. There is a tremendous feeling of achievement and exhilaration once the first thermal flight has been experienced, which never quite disappears, no matter how many long thermal flights are made. This aspect of model flying has its own peculiar challenge, a battle of wits with the elements, and the reward can be literally hours of pleasure.

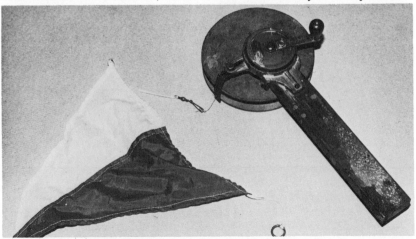

Winch used for storage and rapid retrieval of tow-line after glider tow launch. Pennant and ring can also be seen.

A power pod with small engine fitted provides a practical alternative to tow launching or slope soaring for thermal soarers.

CHAPTER TEN

Conclusion

Continuing success in R/C flying depends on two things, both closely interwoven: good maintenance of equipment and correct attitude. Correct attitude really does include maintenance, but it is simpler to group the points separately. Maintenance begins at the flying field, for after each and every flight the whole model should be checked over, even if the flight was completely uneventful. The routine check should include security of all components of the airframe, security of control surfaces and linkages and condition of the propeller (if fitted).

Repairs and maintenance

My own attitude towards flying field repairs is in general, don't! A hastily carried out repair can fail on the very next flight. An unqualified "don't" is perhaps overstating the case, since minor repairs such as a wheel that has fallen off, a broken control clevis or cockpit canopy that is loose are possible field repairs, but structural repairs are generally inadvisable.

It is all too easy to try and take advantage of the rapid curing properties of either cyanoacrylate or fast epoxy glues, but for success with either of these materials, cleanliness, and often the ability to clamp parts together securely, is a necessity. A poorly executed field repair may make proper repairs difficult when workshop

High performance aerobatic models represent the ultimate for many modellers. Proud builder of this model is Dave Hardacre British National Champion.

89

facilities are available later. If parts are not properly clamped when glued, inaccuracies can immediately be introduced into the airframe; for example, a poorly repaired fin-to-fuselage joint could conceivably result in a model being impossible to control.

Any sign of malfunction of any part of the aircraft/engine/R/C combination must be investigated immediately. Engines should not inexplicably lose power, excessive trim changes should not occur, nor should the servos ever cycle or move unless in response to transmitter control stick movements. This latter happening is known as a 'glitch' and if the model performs any unexpected violent turn, climb or dive, or the throttle is suspected of opening or closing, the reason for the 'glitch' causing the happening should be investigated.

Engine problems usually involve the fuel system or glowplug. Any form of blockage or leak in the fuel feed system will cause the engine to appear to be

Pylon racing is an exciting high speed competitive event, this model built to the International specification.

running too lean. Resist the temptation to open the fuel needle valve to correct the situation, but investigate. Likely causes are dirt in the fuel line, a kinked fuel feed line, a punctured fuel line or the fuel tank clunk weight getting caught up around the vent pipes in the tank. More obscure causes such as a loose carburettor or engine backplate can be checked if the simpler possibilities are not to blame. Glowplug problems are more difficult to diagnose, e.g. an apparently healthy plug may be oxidised to the extent that the effective ignition timing of the engine is altered. Seals around the plug centre pillar can develop leaks which can prevent proper compression of the fuel/air mixture. Glowplugs rarely last the life of the engine; perhaps twenty to thirty runs would be a good average and after this number of runs, unless the element is still looking bright and metallic with no signs of distortion of the coils, it should be replaced. If erratic running or poor idling cannot be traced to a fuel system fault, provided the piston cylinder fit of the engine is good, then a 'tired' glowplug is almost certainly the cause.

R/C equipment maintenance and fault finding are really outside the scope of most R/C modellers, but a few simple points to watch out for can help save an expensive crash. By far the most common areas of failure are in servos. The feedback potentiometer is the major culprit, closely followed by the servo motor. Transmitter control stick potentiometers can also give problems, as can switches, plugs and sockets, in other words all the electro-mechanical areas of the equipment. Any sign of intermittent action of servos should be investigated. It may be that poor switch contact or loose plugs and sockets are to blame, either power supply connections from the battery pack, or individual servo connectors. Switches are best replaced completely, but a gentle 'tweak' can cure a poorly-fitting plug and socket. A good clean with a proprietary contact cleaning/lubricating aerosol can do no harm. Suitable aerosols can be obtained from electronics components stockists.

From time to time it is advisable to examine all leads to servos and batteries etc. for signs of failure. Any plug which is constantly subject to plugging and unplugging should be especially suspect – the aileron servo lead is an obvious one. Look closely at the point where the flexible wire enters the plug or socket, where any sign of breakdown or cracking of the insulation dictates immediate return to the equipment service agent for fitting of a new lead and plug. Receiver aerials need replacing from time to time and should be regularly examined for any sign of damage. Avoid changing crystals too frequently, since the sockets become loose and the silver plating on the crystal pins becomes worn off, possibly spoiling the contact between pin and socket.

Transmitter maintenance can really be confined to cleaning and charging. Use either a small quantity of a strong detergent or methylated spirits to clean the case but be careful not to allow water to enter the transmitter case. Pull the aerial out fully and clean all the sections so that it extends and retracts smoothly.

Attitude towards flying

Attitude towards flying is a more difficult area to advise on. Once the novice pilot has passed the basic take-off circuit and landing hurdles, all too frequently over-confidence results in a wrecked model. Few trainer-type models survive

Make sure that your helper understands what is required and does not release the model until pre-flight checks are completed.

unscathed, but it is a rare novice who will admit to over-confidence, the tendency being to blame the wind, interference, sun glare etc. During the first few months of flying, reaction to minor crises during flights almost always requires a conscious decision, and it is only when control of the model becomes totally instinctive or reflex-activated that the novice becomes an intermediate or experienced flier.

Beware of taking short cuts. The engine cut that would have dictated a landing 100 metres away during early learning flights could be an immaculate 'spot' landing if the controls are the hands of an expert. Ask yourself if the risk is worth it; a walk is better than a wrecked model. Is it really within your capabilities to perform low, slow fly-pasts? They look good, but are your reactions fast enough to prevent a crash in the event of a wind gust tipping the model? At all times seek to fly safely, leave nothing to chance and take your model home in one piece after the flying session.

When basic skills have been mastered, flying skills can be increased enormously by learning basic aerobatic manoeuvres. Most aerobatics are combinations of loops, rolls and spins or 'flicks', and once the basics are learned they can be strung together at will to produce an almost endless variety. Most rudder/elevator models will loop and many will perform quite acceptable rolls, but for full aerobatic potential ailerons are essential.

Before attempting aerobatics be sure of your ability to control the model in any awkward attitude that it may end up in as a result of incorrect control commands whilst learning how to perform the manoeuvre. Even if advanced aerobatics are not felt to be something to be worked towards, the basic Stall Turn and Spin should be attempted, as both stalls and spins are situations that any model can get into, and recovery action should be learnt and practised.

When a model stalls, normally a dive will follow and it is necessary to allow the model to build up full flying speed before recovery into normal level flight is attempted. Once the model has dived and accelerated to flying speed, gentle application of 'up' elevator should bring the model back to level flight. If too much speed has been built up there may be a tendency for the model to 'zoom' towards a

Pupil/Pilot or "buddy box" systems are popular for training – here two UHF (459MH$_z$) systems are seen in use for training.

second stall. Any such tendency must be stopped by gentle application of 'down' elevator before the model starts to climb towards the stall. Spins sometimes occur just as the model stalls. If any rudder offset is present as the models stalls, the wing may drop and the model then starts to rotate rapidly, falling towards the ground with the wings in a fully stalled state.

Recovery from a spin should be automatic; centralise all the controls and the model should recover, probably in a fairly fast steep dive. Pull out of the dive by gentle application of 'up' elevator. Beware of applying too much 'up' elevator when the model is diving steeply, as the stresses produced could break the wings off the model.

If the model continues to spin after the controls are neutralised, it may be necessary to try further measures, and rapidly at that! First feed in 'down' elevator to attempt to produce forward flying speed, which should result in the spin stopping. In extreme cases it may help to apply opposite rudder and open the throttle of the engine up to stop the spin. Once spins and stalls are mastered they make satisfying stunts to include in the aerobatic repertoire.

Future possibilities for R/C model aircraft are many – accurate scale models, racing, pure aerobatic models – once the basic skills of building and flying are mastered the range of activity is limitless.

Do remember that R/C modelling is a hobby and should be fun for you. If any hobby is taken too seriously it can lose much of its appeal, so relax and enjoy one of the greatest hobbies there is.

Third party insurance, if not available through a club or comprehensive household policy, can be obtained through Argus Specialist Publications Ltd, 9 Hall Road, Maylands Wood Estate, Hemel Hempstead, Herts HP2 7BH.

That moment of anticipation! Peter Jackman's 'Monterey' sailplane within milliseconds of release.